Material Culture

Styles, Organization, and Dynamics of Technology

*

A final remark. The snowstorm prevented some participants, including this organizer, and many prospective members of the audience from reaching Detroit and attending the symposium. Thus, for many this publication must serve not only as a record but as a substitute for the symposium. For this reason, the editors and authors are more than usually interested in receiving from readers any comments they may have on the subject and the particular contents of this volume.

Robert S. Merrill

University of Rochester
Rochester, New York
September, 1976

*

Contents

STYLE IN TECHNOLOGY

*

1

Style in Technology—Some Early Thoughts*

HEATHER LECHTMAN

Massachusetts Institute of Technology,
Cambridge, Massachusetts

In 1973, Arthur Steinberg and I suggested that it would be fruitful, particularly for archaeologists who deal almost exclusively with material culture, to investigate technological style as a phenomenon as well as the manner in which individual styles of technology relate to other aspects of the cultures in which they occur.

> In asking what is the cultural component of technology, we are also asking what can technology tell us about culture? We must be concerned not only with the bodies of skill and knowledge of which Merrill (1968)

*Helen Codere, Robert Merrill, Nancy Munn, Cyril Stanley Smith, and Arthur Steinberg read a first draft of this paper. Although they do not all agree with some of my formulations of the issues presented, their criticisms were extremely valuable in the preparation of this brief introduction, and I am most grateful to them for their help.

> speaks, not only with the materials, processes, and products of technology, but also with what technologies express. If we claim that technologies are totally integrated systems that manifest cultural choices and values, what is the nature of that manifestation and how can we "read" it? . . .
>
> . . . We would argue that technologies also [like visual art, music, dance, costume, gesture] are particular sorts of cultural phenomena that reflect cultural preoccupations and that express them in the very style of the technology itself. Our responsibility is to find means by which the form of that expression can be recognized, then to describe and interpret technological style.
>
> (Lechtman and Steinberg In press)

I organized the symposium "Style in Technology" to stimulate interest in the concept of technological style, to see how useful it might be in interpreting cultural data, whether ethnographically or archaeologically assembled, and to elicit some concrete examples of the stylistic component of technology as it functions in specific cultural contexts. The three contributions that follow, those of Marie Jeanne Adams, Mark Leone, and Arthur Steinberg address themselves precisely to this theme.

By style I refer to the formal, extrinsic manifestation of intrinsic pattern. The oft-cited distinction used by linguists between *langue* and *parole* is precisely that distinction between pattern and style. The ordered, redundant phenomena that constitute the patterned structure of all culture are expressed as style in verbal, visual, kinesic and technological behavior. Style is the manifest expression, on the behavioral level, of cultural patterning that is usually neither cognitively known nor even knowable by members of a cultural community except by scientists who may have analysed successfully their own cultural patterns or those of other cultures.

One of the most useful expressions of the relationship between pattern and style is given by Cyril Stanley Smith (In press) who considers style as a phenomenon which is dependent upon structure and is hierarchical in nature.

> Style is the recognition of a quality shared among many things; the quality, however, lies in structure on a smaller scale than that of the things possessing the quality.
>
> . . . The part/whole relationships that produce the externally-visible quality called style seem to be closely analogous to the relationship between internal structure and externally-measurable property that distinguishes one chemical phase from another. . . . Bulk properties of matter

such as density, color, conductivity, crystal structure or vapor pressure by which a chemical phase is identified are not a property of any of the parts, (though they would not exist without them) but rather are external characteristics depending on the *pattern of interaction* between the atomic nuclei, electrons and energy quanta and the extension of this pattern by repetition throughout the entire volume of the phase concerned. (italics mine)

. . . Style is hierarchical; it resides at all levels, or rather between any inter-relatable levels.

Two features of style emerge from Smith's model of the structural nature of all systems, whether physical, biological, or social: 1) within any system "it is the relationships of communication, not the parts themselves, that lock in to reinforce and stabilize a larger pattern . . ." and it is our perception of the formal arrangement of those interrelationships that we recognize as style; 2) the particular patterns of relationships among interacting parts are different at different levels within the system, thus style is hierarchical and its manifestation depends upon where we locate to observe the interactions.

For example, a crystal exists at one level of structural hierarchy higher than that of the ordered arrays of atoms which comprise it. Crystallinity, as an identifiable property, represents a physical style of matter. It is a style dependent upon the *repetition* of the local symmetries of the unit cell throughout the crystal. Though the essence of crystallinity lies in the unit cell, the crystal is the extension of this. The physicist who studies the internal structure of the atom does not see the unit cell, and the crystallographer who studies the symmetries of the unit cell does not necessarily see the crystal as defined by their extension and the limits of their extension. Different styles become observable at different levels of the hierarchy of aggregation. (Smith In press). The properties identifying crystal grains in the aggregates studied by the metallurgist or the polyhedral crystals studied externally by the traditional mineralogist represent a new state of things, a style emerging at a higher level from the patterns of atomic interaction at lower levels.

Scholars rarely experience any difficulty in talking about style or in describing its formal elements when we are considering objects, that is, the physical products of certain types of behavior. The large majority of studies in art history are predicated upon the scholar's ability to group works of art by their formal, stylistic properties. Archaeology is similarly dependent upon the use of stylistic categories of artifacts which are derived from study of their formal characteristics. What we haven't seemed to recognize or at least paid much attention to is that the activities themselves which produce the artifacts are stylistic. "Material culture is the name given to the man-made physical products of human behavior patterns . . ." (R. Spier 1970:14); and it is precisely those behavior patterns that constitute the style of technology. Technological behavior is characterized by the many elements that make up technological activities—for example, by technical modes of operation, attitudes towards materials, some specific organization of labor, ritual observances—elements which are unified nonrandomly in a complex of formal relationships. It is the format or "package" defined by these relationships that is stylistic in nature, and it is the style of such behavior, not only the rules by which any of its constituent activities is governed, that is learned and transmitted through time.

Leone's analysis (1973) of Mormon town plans and fences as well as his discussion of Mormon "sacred technology" in his symposium paper here (pp. 87–107) are excellent illustrations of technological style. He sets out the characteristic web of interactions of certain "pieces" of Mormon technology with the culture's ideology, religion, and social structure and provides the principle he believes underlies those interactions. That principle is "mutually exclusive compartmentalization: the use of categories whose closest members have no contact with each other" (see p.). The building of fences —fences that separated house from house and therefore family from family, sheep and goats from horses and cattle, natural predators from everything domesticated, and wind-borne sands from the farmyard and garden (1973:143–4)—was an ecological necessity in the semi-arid Utah plains, in that fences ". . . separate[d] competitive niches from each other and protect[ed] all artificially created niches from the universally destructive wind" (1973:144). But the building of fences was important also because behind his fence a Mormon could raise a garden, could make the desert bloom, could demonstrate his worth before his God. "The state of a man's yard is the state of his religion" is the way one Mormon expressed the relationship (Leone 1973:147). Fences were crucial for subsistence but also were fundamental to religious and social life. Leone argues further that the physical compartmentation of the Mormons' world, the parceling out of that world through the technology of fence building, created the kind of cultural environment in which the individual grew up and that that environment must have had a cognitive effect upon him. "This environment is a result of the way Mormons have had to think about their world and of what the Mormon idea-system was and has become. If fences are a piece of enabling . . . technology, then it is reasonable to suppose that the technology enables them to think in certain ways, as well as to grow crops in certain ways. As a result, this system of technology should have cognitive consequences" (1973:147). Leone goes on to describe what some of those consequences are and concludes that ". . . fences enable them [Mormons] to redeem the earth and manipulate and act out the categories used to deal with the world" (1973:148), categories that are essentially exclusive and incompatible. The technological style observed in Mormon fencing and town planning is an expression, on the level of technological behavior, of an underlying cultural pattern of "mutually exclusive compartmentalization." The technology incorporates and transmits the principle itself.

Style can be thought of, then, as the sensible manifestation of pattern; and technological style is expressed "emic" behavior based upon primarily "etic" phenomena of nature: copper melts at 1083°C and, after cold working, recrystallizes at about 300°C, but why a culture technically capable both of casting and of forging copper elects one of these manufacturing techniques to the exclusion of the other is not explained by these properties of the metal. If, through appropriate study, we can describe the elements of any given technological style and can determine the relationships among them, in other words if we can successfully define the style, what can we then say about the intrinsic cultural pattern or patterns of which it is an expression? This issue is vital to archaeological research, for the single subsystem of a once-living culture that archaeologists can reconstruct and understand almost in its entirety is the technological subsystem. Binford (1962:218) has observed, "It has often been suggested that we cannot dig up a social system or ideology. Granted we cannot excavate a kinship terminology or a philosophy, but we can and do excavate the material items which functioned together with these more behavioral elements within the appropriate subsystems." But what we can and do excavate are technologies. Furthermore, it is wholly within our capability to determine accurately the technical events that went into the manufacture of the kinds of items to which Binford refers, from the gathering of the natural resources through the various stages of processing, alteration, and final rendering of the artifact. These events are all behavioral, and they proceed in a formal way. I am calling technological style that which arises from the formal integration of these behavioral events. It is recognizable by virtue of its repetition which allows us to see the underlying similarities in the formal arrangement of the patterns of events.

An example to illustrate the argument I have been developing might be helpful. It comes from my work in the field of pre-Columbian metallurgy. Although I have presented the data and the issues surrounding them before (Lechtman In press 1971, 1973), the problems raised remain to be solved, and the case seems an instructive one for the present discussion.

When one considers the development of metallurgy in the Andean area prior to the Spanish invasion, one is struck by the nature of the alloy systems that dominated the course of Andean metallurgy for at least two millennia. They are remarkable when compared with the alloys used by other ancient culture areas that practiced sophisticated metallurgy, for example, the Near East, Europe, China, both because they are so different from those others and because they appeared on and were a primary stimulus to the Andean metallurgical scene long before alloys of bronze were produced by Andean metalworkers. Whereas the other great prehistoric traditions of metallurgy were founded upon the utilization of copper and, very early, of copper-arsenic and/or copper-tin bronze, the production of bronze was a relatively late development in Andean metallurgy. It was preceded by two thousand years of experience, at a high level of technical sophistication, with a variety of copper-based alloys that were developed for the colors they would impart to the artifacts made from them. These deliberate alloys (as distinct from naturally occurring ones), of copper-silver, copper-gold (usually referred to as tumbaga), and copper-silver-gold, were not exclusive to Andean metalworkers. They were utilized by other peoples prehistorically, but only occasionally, and nowhere else did they form the backbone of the indigenous metallurgy, influencing markedly the course of its own internal development.

The colors the early Andean metalworkers wanted to achieve were the colors of silver and gold. When objects were made directly from either of these precious metals, the color was attained automatically. But objects made from other metals—copper, for example—could be given the desired appearance by providing them with a surface layer of metallic silver or gold. The usual way of gilding or silvering metal objects, as such surface coating is called, the method utilized by the ancient metallurgies of the Old World, is to apply a thin layer of the metal directly to the object's surface. For example, gold in the form of thin foil or leaf, in the form of a fine powder, in molten form, or as an amalgam with mercury, was applied to metal objects to provide them with surfaces of gold.

The Andean process for achieving the same end is almost the reverse of that just described. Rather than placing the precious metal upon the surface of the object, as an added element, nonintegral with the object itself, they placed the precious metal within the bulk of the object by incorporating it as one of the constituents of the original alloy from which the object was later fabricated. For example, an alloy of 80 percent copper and 20 percent gold contains the gold in solid solution, distributed homogeneously throughout the alloy. The alloy does not look golden; rather it is a distinctly coppery color, and an object made of such an alloy would appear as copper. But the gold is present internally and can be made manifest at the object's surface by treating the surface so that the copper there is removed selectively, leaving the gold in place. The gold is "developed" at the surface by eliminating the other components of the alloy—copper or minor constituents such as silver or lead—at the surface and leaving only the gold coating layer. The Andean peoples devised various alloy systems and associated techniques for chemically treating the surfaces of those alloys in order to obtain gold- and silver-looking objects. The first such alloy seems to have been a simple copper-silver binary system which gave a silver surface to objects made from it. Tumbaga then appeared (copper-gold) and eventually the ternary copper-silver-gold alloy that is so characteristic of north Andean metallurgy as practiced by the Kingdom of Chimor whose master goldsmiths were brought to Cuzco by the conquering Inca to work for the royal lineage.

The principle of incorporating within a metal object the constituent which was later to become its most important external quality was the governing idea that stimulated the invention of a whole group of alloy systems in ancient Peru. The idea spread from Peru to the peoples of Colombia where it was adopted and adapted to a technical tradition of handling metal that was almost diametrically opposed to the Peruvian tradition (casting in Colombia vs. forging in Peru). It was an idea to which Andean metallurgy was committed for the major part of its history and through which some of its most inventive developments took place. These considerations suggest that the principle itself was a powerfully motivating one. Its application resulted in a metallurgical style unique to the Andes.

This metallurgical style, based upon processes of surface enrichment and depletion, has been identified through laboratory studies on large numbers of metal objects from the Andean culture area. The majority of these were used for ceremonial or political purposes—as funerary masks applied to mummy bundles, large expanses of metal sheet that lined the interiors of palace and temple walls, small plaques sewn to garments for status display, figurines, decorative vessels, and so forth. That gold and silver should have been the colors desired is consistent with the obviously fundamental role those metals played in Andean cosmology and statecraft. Our best evidence is from the Late Horizon (ca. 1470–1532 A.D.) when gold and silver were the exclusive property of the Inca, that is, of the state. The royal lineage was believed to have issued directly from the sun, the first Inca being the son of the sun. Gold was considered the "sweat of the sun," silver the "tears of the moon." Ritual performed for the sun cult was one of the most important of the priestly functions. But neither the weaving of these materials throughout the religious and cosmological orders nor their political manipulation accounts for the particular metallurgical style that was the Andean response to a desire for gold- or silver-looking objects. Nor, for that matter, do any of the other explanations that have been offered, particularly for the widespread use of the copper-gold alloy tumbaga, account for the evolution of the technological style. William Root (1951) argued that although tumbaga may have been used because it was easier to cast such an alloy than it was to cast copper; it made objects harder because copper-gold alloys are harder than either copper or gold alone; it economized on gold because a small amount of gold within the alloy was sufficient to create a gold surface and, therefore, a "gold" object; the most likely reason for its use was simply that those who made and used the alloy preferred the color of gold to that of copper. Even though the constellation of tumbaga may have been composed of some or all of these factors, they seem to me to fall short of explaining what appears to have been an overwhelmingly central tendency, a pattern of Andean metallurgical development. We can recognize the technological style. But what does it express? What gave it the tenacity it most certainly had such that Andean metalworkers seem to have been locked into a style of operating, though with evident opportunity, within that convention, to be inventive.

I have suggested (Lechtman In press) that what lay behind the technological style were attitudes of artisans towards the materials they used, attitudes of cultural communities towards the nature of the technological events themselves, and the objects resulting from them. The basis of Andean surface enrichment systems lies in the incorporation of the essential ingredient into the very body of the object. The essence had to be present though invisible. The style is a playing out of the notion that:

> . . . the essence of the object, that which appears superficially to be true of it, must also be inside it. The object is not that object unless in contains within it the essential quality, even if the essence is only minimally present. For without the incorporation of the essence, its visual manifestation is impossible. . . . Although ideological considerations may have had little to do with the innitial working out of [the technical] procedures, I feel sure that the way in which the Andean peoples perceived such processes or at least the objects that resulted from their use had a great deal to do with the way in which the technology emerged and matured. Belief systems and attitudes toward materials supported the technology and gave rise to further developments along similar lines (Lechtman In press.).

The technological performance was supported by a set of underlying values. I have suggested what I think one of those values or standards may have been, as I interpret it from that portion of the performance which is purely technical, the events of production that remain part of the physical structure of the object.

Examples such as this one indicate that it is possible to determine the technological styles that underlay particular sets of artifacts of archaeological study. The much more difficult step is to argue from a confident understanding of the style of technological behavior to more fundamental, deeper cultural patterns which informed that behavior. Yet, if we could do that, archaeology would have a powerful tool for getting closer to those aspects of past cultures that cannot be determined directly—for example, the realm of ideology, values, philosophy—yet whose imprint should be accessible through the study of behavior as it is observed in the material record. "The objects man has learned to make are traditionally termed material culture. Culture is intellectual, rational, and abstract; it cannot be material, but material can be cultural and "material culture" embraces those segments of human learning which provide a person with plans, methods, and reasons for producing things which can be seen and touched" (Glassie 1968:2). What are our chances for success in determining the "plans" and "reasons" behind technological behavior as those are revealed in the artifacts themselves?

In the example of Andean metallurgy just cited, how do I know that my interpretation is correct, as plausible as it may seem from my frame of reference? Does my ethnocentrism come between me and the data? And even if we can get at the standards *for* technological behavior by looking at patterns *of* technological behavior (see Keesing 1969:208), what is there to suggest that those same standards inform any of the other interacting subsystems of culture?

The view of culture as cognitive code which is separate and distinct from material artifacts and behavior has been most forcefully represented by Goodenough. In fact, Keesing (1969:207) considers the ". . . effective theoretical separation between cultural codes—cognitively based normative systems—and their enactment in behavior . . ." one of the major breakthroughs in social anthropology of recent years. In essence, Goodenough (1964) argues that there are two domains in culture, the ideational and the phenomenal, the latter being an artifact of the former. He states that the phenomenal order of events, of behavior, of artifacts within a human community:

> . . . exhibits the statistical patterns characteristic of internally stable systems, as with homeostasis in the living organism. Similar, but never identical, events occur over and over again and are therefore isolable as types of event and patterned arrangement. Certain types of arrangement tend to persist and others to appear and reappear in fixed sequences. An observer can perceive this kind of statistical patterning in a community without any knowledge whatever of the ideas, beliefs, values, and principles of action of

> the community's members, the ideational order. . . . The ideational order, unlike the statistical order, is nonmaterial, being composed of ideal forms as they exist in people's minds, propositions about their interrelationships, preference ratings regarding them, and recipes for their mutual ordering as means to desired ends (Goodenough 1964:12).

Whether or not one subscribes to this view, I would argue, on the basis of Smith's model, that the patterns of behavior we see depend upon the level at which we isolate them for study and at which they exhibit style. To the extent that style is hierarchical, the scale of resolution at which we can study behavior may not be the same as the scale at which ideation operates, but one is a manifestation of the other and there must be a structural connection between the two. While it may be extremely difficult to arrive at the underlying structure in culture below the level at which we perceive technological style, for example, the persistent attributes of the style relate to a formal arrangement of operations and that arrangement, in itself, carries a heavy load of meaning. Communication among the parts of a system are the backbone of style, but the style as a whole, once perceived, is itself a form of communication.

Despite the primacy of language as *the* human means of communication and the *sine qua non* of culture, no one would argue that it is the sole device for the sharing of socially acquired knowledge nor that verbal communication functions best in transmitting all varieties of messages. As Frake has pointed out, what we want to do is both to describe socially meaningful behavior and to discover the rules behind such behavior. The entire domain of socially interpretable acts and artifacts, that is, the total domain of "messages" is the concern of ethnography which "seeks to describe an infinite set of variable messages as manifestations of a finite shared code, the code being a set of rules for the socially appropriate construction and interpretation of messages" (Frake 1964:132). Implicit in the equation of "socially interpretable acts and artifacts" with "messages" is the understanding that a shared cultural code is expressed along a variety of communication channels, among which are reckoned acts of behavior and artifacts. In that case, archaeology can address itself to at least some of the behavioral and all of the material elements which make up the total domain of messages within a community. Artifacts are the products of appropriate cultural performance, and technological activities constitute one mode of such performance. What I have called the style of technological behavior is the rendering of appropriate technological performance. The style itself is the rendition, and the measure of its appropriateness, as determined archaeologically at least, lies in its reiteration. Furthermore, it is the synthesizing action of the style, the rendering of the performance, that constitutes the cultural message. Technologies are performances; they are communicative systems, and their styles are the symbols through which communication occurs. The relationships among the formal elements of the technology establish its style, which in turn becomes the basis of a message on a larger scale.

I would argue further that what is communicated through the expression of technological style is not communicable verbally, despite Frake's skepticism that "... it [is] difficult to conceive of any act, object, or event which can be described as a *cultural* artifact, a manifestation of a code, without some reference to the way people talk about it" (1964:133). Bateson (1972:137) credits Anthony Forge for the remark by Isadora Duncan: "If I could tell you what it meant, there would be no point in dancing it." Recent studies of indigenous art are providing increasing evidence for the use of nonverbal systems of expression to communicate fundamental ideas about the natural and social order. In his studies of Abelam flat painting, Forge (1973) has indicated that some of the messages the art carries do not seem to operate at a conscious level, but rather that there is a "grammar" of painting which is as unconscious as is the grammar of spoken language. The styles of the painting, he argues, are systems of meaning, systems of communication which "... unlike those to which we are used, exist(s) and operate(s) because [they are] not verbalized and probably not verbalizable, [they] communicate(s) only to those socialized to receive [them]" (1973:191). Similarly, Nancy Munn has documented, in a series of provocative articles (1964, 1966, 1973), the method by which Walbiri communicate principles of cosmic order through their spatial arrangements, in media of two and three dimensions, of the traditional elements of their visual vocabulary. She points out, however (1973:216, note 1), that such iconographic systems may be given somatic form in dance or the enactment of ritual. I do not think it far fetched to suggest that part of the communicative aspect of technologies lies in the somatic nature of their performance which involves not only the articulation of body and tool or body and material but the exemplification of skill. The case is stated elegantly by Hallowell (1968:235).

> Systems of extrinsic symbolization necessitate the use of material media which can function as vehicles for the communication of meanings. Abstraction and conceptualization are required since objects or events are introduced into the perceptual field as *symbols,* not in their concrete reality. Thus systems of extrinsic symbolization involve the operation of the representative principle on a more complex level than do processes of intrinsic symbolization. In case of *Homo sapiens,* extrinsic symbolic systems, functioning through vocal, graphic, plastic, gestural, or other media, make it

> possible for groups of human beings to share a common world of meanings and values. A cultural mode of adaptation is unthinkable without systems of extrinsic symbolization.

Technologies are such symbolic systems. A good example in support of this argument is given by M. J. Adams' work (1971, 1973) on the island of East Sumba, Indonesia. Dyed textiles are not only the major "visual art" of the island but are also exceptionally important as ceremonial costume, as wealth in ritualized gift exchange, and as sacred objects (1971:322). Adams describes the technical procedures and work schedules that are associated with the production of textiles, from the planting of the cotton seeds to the final weaving of the ikat designs into the cloth. At each stage of production she shows the close relationships between the activities undertaken in the work routine and phases in the Sumbanese life cycle, especially those which relate to the physical and social maturation of women. "In myth, ritual and social rules on Sumba, the stages of textile work are consistently linked to the progressive development of individual human life. These stages provide an overarching metaphor for the phases of the Sumbanese life cycle" (1971:322). It is not only that, as Adams argues, the procedures and schedules of work provide metaphoric schema for other symbolic systems, but that the technological acts themselves constitute a symbolic system.

There is no question, then, that we can excavate artifacts and reconstruct the technologies behind them. In doing so, we may discover specific technological styles which are renderings of appropriate technological behavior communicated through performance. The culturally accepted rules of the performance are embodied in the events that led to the production of the artifact. We should be able to "read" those events, if not all of them at least those of a technical nature, by laboratory study of the materials that make up the artifacts in question. The history of the manipulation of those materials is locked into their physical and chemical structure; the methods of materials science can interpret that technical history.

Having come thus far, what remains is the task of describing the relational order between the symbolic, technological events, and that which they symbolize—of coming to grips with decoding the technological system of communication. The interpretation of symbolic content in archaeological data is extremely difficult. We can rely upon the fact that the formal relationships that exist in any iconographic scheme or that constitute a technological style are rarely if ever dictated solely by the environment. They largely reflect cultural choices. That a particular community tills in a certain way or pots in another or builds in yet another is certainly affected by the nature of the soil and micro-climate, the clay, or the building materials available. But those are immutable conditions in and around which people elaborate technological behavior along lines that are meaningful socially, economically, and ideologically. The rules behind their choices are what we are after.

Although I have been talking about "technological style" as a phenomenon, that should not imply that any given cultural community is characterized by only one such style. In fact, several styles may operate synchronically, each having developed as it did as a result of a multitude of factors including the nature of the technological "task" itself (the building of irrigation ditches as opposed to the construction of a ceremonial dance mask), the social group performing the technological activity or for whom it is performed (e.g. commoner/elite, peasant/landlord, men/women), the cultural subsystem in which the technological events primarily operate (social, technological, ideological), the properties of the environment being manipulated by the technology, and so on. As Binford has argued, the artifacts one studies or the class of items they represent ". . . are articulated differently within an integrated cultural system, hence the pertinent variables with which each is articulated, and exhibit concomitant variation are different" (1962:219). Technological behavior is manifest in all activities in which the natural or social environment is directly manipulated, but the style of that behavior may be different according to the particular integration of the technological complex within any given subsystem of the total cultural scene. Styles for the production of mundane goods may be different from styles for the manufacture of sacred objects; elite styles may be different from folk styles. In attempting to decode the message carried by technological style we must be cognizant that the message may not be the same for each style encountered within a given cultural community. The intriguing questions are: when the styles *are* the same, when the message *is* reiterated, a) what is the message; and b) what are the socio-cultural circumstances that stimulate styles which bear similar messages? For example, will we tend to find that, despite the obvious differences in technique between women who weave sacred or ritual garments and men who cast ritual vessels for use in the same ceremonies as the cloth, the styles in which these otherwise disparate technologies operate are based on similar underlying patterns of technological behavior because what each expresses has to do with what the ritual expresses? Will technologies within the ideological subsystem tend to be stylistically alike because of the relationships they bear to the underlying ideology? Or are we more likely to find that we cannot always apply the same sorts of categories Binford (1962) uses for artifacts to technologies, that is, technomic, sociotechnic, ideotechnic categories based on the primary functional context of the artifact or, in this case, of the behavior? My guess is that technological styles will appear similar wherever the message they carry relates to idea systems, values, orientations that cross-cut the social, technological, and ideological realms of a society. In fact, archaeological identification of similar technological styles within these various subsystems should point to a message widely expressed throughout the culture and, perhaps, give us a better handle on what that message may have been, of how to reconstruct that portion of the cultural code which is manifest in the style. Perhaps both specific styles relating to specific cultural spheres and intracultural styles exhibiting a similar expression in many of those spheres will prove to be characteristic of styles in technology.

Returning to my earlier question, if we allow ourselves to interpret the meaning of technological style once we have defined the style, how do we know that our interpretations are correct? If it is true that technological styles are both meaningful in themselves and are manifestations of cultural codes, then a style which seems indigenous (as opposed to introduced from the outside), persistent, and stimulative, may have served as the model for the expression of "message" in other media or in other subsystems of the culture than those in which it is first observed and for which it seems particularly characteristic (as, for example, the style of metallurgical technology). The problem for the archaeologist is where to look for the evidence. Does one begin with another technology, one which appears of similar importance and which one suspects was designed in part to bear ideational content (in the Andes this would unquestionably be the production of cloth) to see if the style of that system is organized around a similar model and is expressive of similar preoccupations? Will we find, as in the case of Andean metallurgy, that the model applies primarily to behavior the products of which, the artifacts, are primarily operative in the social or ideological subsystems of the total cultural system? Does *cumpi* cloth, the finest textiles woven for the Inca by the specially chosen and trained *aclla* female weavers or the male *cumbicamayocs,* display in its structure or the manipulation of its materials the same patterns of formal relationships that underlie the style of the royal metallurgy? Or does it make more sense to look not at another creative technological complex but rather at a different system of cummunication, one which relies upon technological input but whose focus is elsewhere, the realm of "art," for example? If we were to seek the aesthetic locus of a culture, in the sense that Maquet (1971) uses that concept, and were to investigate the symbolic and expressive content of that locus or elements within it through study of its remaining artifacts, would we find similar patterns indicative of ideas such as the incorporation of essences, reiterated in the locus structure? Testing hypotheses that have to do with the message content of sets of artifacts is exceedingly difficult, but I think we must make the attempt if we expect to make any headway in understanding the interplay between ideas and performance in the technological sphere of life. That we must proceed cautiously, avoiding the obvious pitfalls, has been amply stressed by others as well (e.g. in the work of Friedrich 1970 and White and Thomas 1972).

I have dwelt at some length with the concept of style in technology as that concept applies to archaeological situations, because archaeology must constantly explore new strategies for mining its artifacts for all that they are worth. My feeling is that systems of technology, as reconstructed primarily from evidence provided by laboratory study of artifacts, are worth more than we have sought from them. I am suggesting that one tactic we might exploit is the study of technologies as systems that proceed in a stylistic manner, some of the elements of which we can determine with little error. Defining the parameters of a particular style may help in eliciting from the technology information about its own symbolic message, and about cultural code, values, standards, and rules that underlay the technological performance. It is obvious that the ideas I have set forth here must be tested through ethnographic fieldwork. That may not be easy, at least in the case of many of the traditional societies whose technologies have undergone more rapid change than other aspects of life as a result of Western economic imperialism. When the technology involved is a Western imposition or import, there is little reason to suspect that any of the traditional sets of values still inform modern technological behavior. But we might still find clues in those areas of life that were not central to Western development schemes, in the traditional arts, for example. If systems of beliefs are reflected in objects of art, they ought also to be reflected in the processes by which art objects are produced. Perhaps it is in the technology of art that we might look for evidence of the symbolic content and code-bearing nature of technology. On the other hand, we ought also to investigate the technologies of modern, industrial societies where, although the data may be more complex and difficult to assess because of our closeness to it, we may have a better chance of observing the kinds of relationships I have tried to define.

I would encourage anthropologists—whether they practice ethnography or archaeology—to reconsider the richness of technological behavior and to explore that behavior not only as moderator between society and the natural world but as an important vehicle for creating and maintaining a symbolically meaningful environment. The maintaining of particular technological styles has probably always been one of the effective ways by which communities have enculturated values through nonverbal behavior.

LITERATURE CITED

Adams, Marie Jeanne

1973 Structural Aspects of a Village Art. American Anthropologist 75: 265–279.

1971 Work Patterns and Symbolic Structures in a Village Culture, East Sumba, Indonesia. Southeast Asia 1:321–334.

Bateson, Gregory

1972 Style, Grace, and Information in Primitive Art. *In* Steps to an Ecology of Mind. G. Bateson, ed. Pp. 128–156. San Francisco: Chandler.

Binford, Lewis R.

1962 Archaeology as Anthropology. American Antiquity 28:217–225.

Friedrich, Margaret Hardin

1970 Design Structure and Social Interaction: Archaeological Implications of an Ethnographic Analysis. American Antiquity 35:332–343.

Forge, Anthony

1973 Style and Meaning in Sepik Art. *In* Primitive Art and Society. A. Forge, ed. Pp. 169–192. New York: Oxford University Press.

Frake, Charles O.

1964 Notes on Queries in Ethnography. *In* Transcultural Studies in Cognition. A. K. Romney and R. G. D'Andrade, eds. American Anthropologist 66, Part 2, No. 3:132–145.

Glassie, Henry

1968 Patterns in the Material Folk Culture of the Eastern United States. Philadelphia: University of Pennsylvania Press.

Goodenough, Ward H.

1964 Introduction. *In* Explorations in Cultural Anthropology. W. H. Goodenough, ed. Pp.1–24. New York: McGraw-Hill.

Hallowell, A. Irving

1968 Self, Society, and Culture in Phylogenetic Perspective. *In* Culture—Man's Adaptive Dimension. M. F. Ashley Montagu, ed. Pp. 197–261. New York: Oxford University Press.

Keesing, Roger M.

1969 On Quibblings Over Squabblings of Siblings: New Perspectives on Kin Terms and Role Behavior. Southwestern Journal of Anthropology 25:207–227.

Lechtman, Heather
In press Issues in Andean Metallurgy. *In* Pre-Columbian Metallurgy of South America. E. P. Benson, ed. Washington, D.C.: Dumbarton Oaks.
1973 The Gilding of Metals in Pre-Columbian Peru. *In* Application of Science in Examination of Works of Art. W. J. Young, ed. Pp. 38–52. Boston: Museum of Fine Arts.
1971 Ancient Methods of Gilding Silver—Examples from the Old and the New Worlds. *In* Science and Archaeology. R. H. Brill, ed. Pp. 2–30. Cambridge, Mass.: M.I.T. Press.

Lechtman, Heather and Arthur Steinberg
In press The History of Technology: An Anthropological Point of View. *In* Proceedings of the International Symposium on the History and Philosophy of Technology. University of Illinois at Chicago Circle, 1973.

Leone, Mark P.
1973 Archeology as the Science of Technology: Mormon Town Plans and Fences. *In* Research and Theory in Current Archeology. Charles L. Redman, ed. Pp. 125–150. New York: John Wiley.

Maquet, Jacques
1971 Introduction to Aesthetic Anthropology. Current Topics in Anthropology 1, Module 4:1–38.

Merrill, Robert S.
1968 The Study of Technology. *In* International Encyclopedia of the Social Sciences. David L. Sills, ed. Vol. 15:576–589. New York: Macmillan.

Munn, Nancy D.
1973 The Spatial Presentation of Cosmic Order in Walbiri Iconography. *In* Primitive Art and Society. A. Forge, ed. Pp. 193–220. New York: Oxford University Press.
1966 Visual Categories: An Approach to the Study of Represenational Systems. American Anthropologist 68:936–950.
1964 Totemic Designs and Group Continuity in Walbiri Cosmology. *In* Aborigines Now. M. Reay, ed. Pp. 83–100. Sydney: Angus and Robertson.

Root, William C.
1951 Gold-Copper Alloys in Ancient America. J. of Chemical Ed. 28:76–78.

Smith, Cyril Stanley
In press Structural Hierarchy in Science, Art, and History. *In* On Aesthetics in Science. Judith Wechsler, ed. Cambridge, Mass.: M.I.T. Press.

Spier, Robert F. G.
1970 From the Hand of Man. Boston: Houghton-Mifflin.

White, J. P. and D. H. Thomas
1972 What Mean These Stones? Ethno-taxonomic Models and Archaeological Interpretations in the New Guinea Highlands. *In* Models in Archaeology. D. L. Clarke, ed. Pp. 275–308. London: Methuen.

2

Style in Southeast Asian Materials Processing: Some Implications for Ritual and Art*

MARIE JEANNE ADAMS

Harvard University, Cambridge, Massachusetts

At a time when various scholars are urging renewed interest in material culture studies (Spier 1970, Mori 1972, Oswalt 1973, Richardson 1974, Lechtman and Steinberg In press), I am calling attention to one especially neglected aspect of the ethnographic record, that is, the technical processing of materials. To suggest the potential of studies in this area I will point to similarities, in certain selected cases in Southeast Asia, between the manner of processing materials and the nature of artistic efforts and ritual sequence. Specifically, I shall indicate similarities be-

*I would like to thank Heather Lechtman for her editorial help.

tween the widespread and skillful manipulation of cloth fibers and the emphasis on binding rituals, and between the staged decomposition of vegetal substances and the formal sequences of funeral ritual.

This exploratory effort represents but a fraction of what should and could be done on this subject and only a small step toward the comprehensive examination that Lechtman and Steinberg (In press) call for in presenting their concept of style in technology.

Possible links between favored methods of materials processing and other characteristic features of Southeast Asian culture, particularly in the realm of art and ritual, emerged for me from the pages of ethnography, from my visits to the Southeast Asian mainland, and from my fieldwork in villages in Eastern Indonesia. Although I can deal with the subject only on a very broad level here, I am encouraged to offer this paper for two reasons: as a stimulant to observation of materials processing by others planning research and fieldwork, and as a way of obtaining refinements, qualifications, and perhaps new insights from an audience interested in material culture.

Discussions of the peoples of Southeast Asia usually point to two major cultural groupings during the past two hundred years, distinguishing between the rice cultivators of the plains who practice Buddhism and the horticulturists of the highlands who adhere to spirit cults. In the lowlands social hierarchy and monumental centers for royalty and religion provide further contrasts to the "egalitarian" and humble character of hill-village communities (for area map, see Fig. 1). In spite of these obvious differences, compounded by a variety of ethnic and linguistic groupings (see Kunstadter 1967), ethnographers in the past (Heine-Geldern 1923; Bacon 1946) have noted a number of traits which recur with sufficient frequency to be considered characteristic: concrete features such as houses on piles; colored sashes and tunic as costume; personal body alterations such as blackening teeth and tattooing; practices like the use of fish but not milk products, the chewing of betel nut, the fermentation of foods and the use of poisons for fish-kills and on weapons.

Although such trait lists are uneven and unsatisfactory for conveying a general comprehension of the regional way of life, nevertheless I share an interest in looking at broadly distributed features. By examining processing methods I hope to discern some common and basic behavioral characteristics that, so to speak, "go below" the level of the dual divisions outlined above. My aim is to select examples of technical processes that stand out for frequency and elaboration of application in order to demonstrate the quality of similarity between these and other distinctive features of Southeast Asian culture. In this exploratory effort, the evidence I offer for similarity is not drawn from the ideological sphere (although much could be provided from this source) but from the field of observable behavior or products.

One basic feature of Southeast Asian culture has been observed in concrete ways by many and summarized succinctly by Loofs (1964): technical problems are solved with plants. These solutions involve major elements of material culture such as housing, clothing, and tools. Southeast Asia has been called a botanical paradise because of the presence of a great variety of plants: flowers, fruits, medicinal flora, essences, spices, and so on. The traditional societies reflect the cumulative interest of previous generations in exploiting that environment. Characteristic of this exploitation is a special emphasis on complex processing sequences that

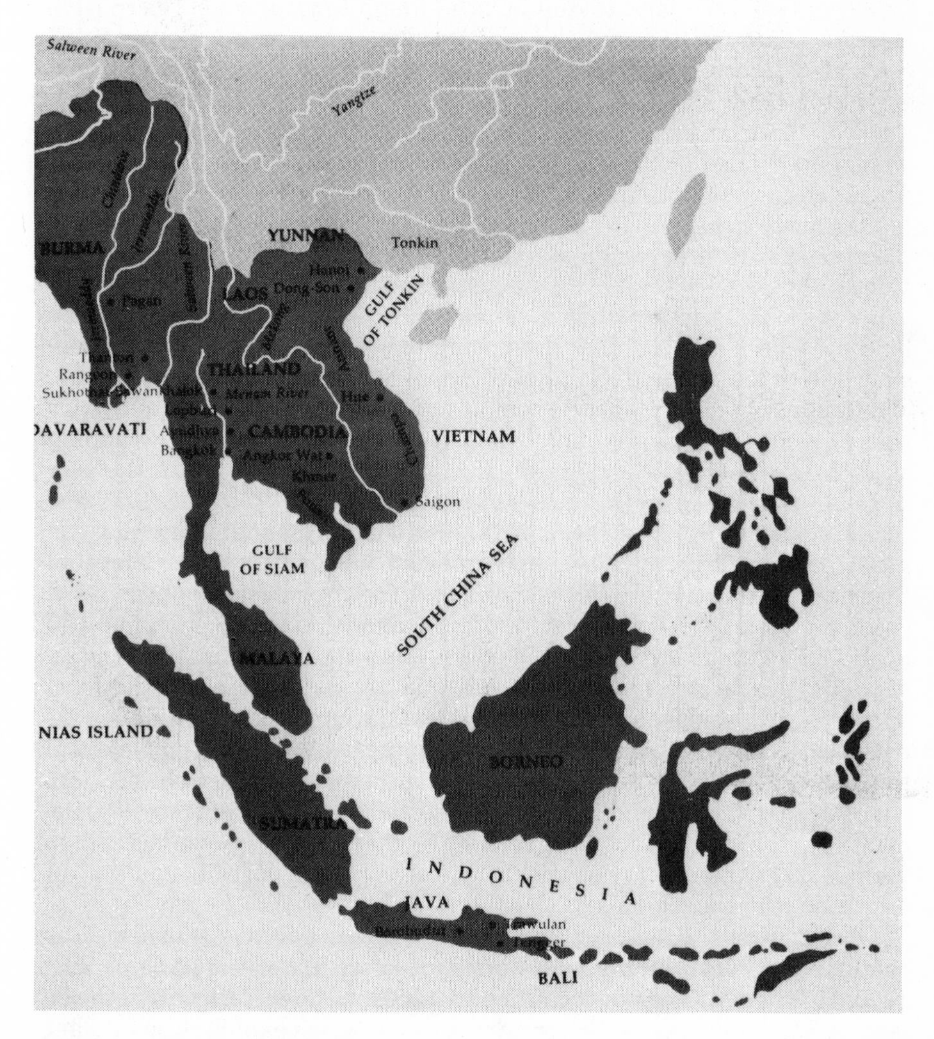

Figure 1. Map of Southeast Asia.

make vegetal products edible or usable (Pelzer 1945). Elaborate techniques have been developed for modifying poisonous vegetables and for preserving food. The complex exploitation of plants that produce fibers suitable for binding is yet another area in which great elaboration has occurred. An interesting example is provided by Bernatzik (1970:620) in his description of procedures used by the Hmong (Miao) people of North Thailand to treat hemp, a widely used textile fiber in the highlands of northern Southeast Asia and South China. Hemp stalks are soaked in water until rotten. They are dried and the woody parts are removed by breaking the stalks and taking out the tough inner tissue. Fibers from this tissue are twisted by hand into long threads and wound on sticks. Several of these strands are spun together on a spinning wheel and wound on large spindles. The yarn then goes through three boiling sequences: first boiled in water and potash, then rinsed and unwound; rewound and boiled in lye of wood ash, then rinsed and tested; and finally for the third time rewound into hanks for boiling with beeswax. The yarn is then ready to be used for weaving. These procedures illustrate the complex processing of vegetal material that is part of Southeast Asian traditions.

Before looking in detail at a similarity I am suggesting between the binding aspect of cloth-fiber processing and binding rituals, let us consider the broad frame of Southeast Asian culture within which binding and pounding are frequently employed techniques for dealing with materials in general. By binding I mean both tying and wrapping around, like lashing, which are the first two meanings given for the verb to bind in the *Concise Oxford English Dictionary* (1971:217). Like features in the old-fashioned trait lists, binding and pounding are not emphasized with uniform consistency in every village nor in every level of Southeast Asian society. In some places these processes appear, as if "stored" in lesser, peripheral ways. In many, however, binding and pounding methods are elaborated and prominent, and their emphasis recurs sufficiently over great distances that these processes should form part of our conscious image of the region. Of course these techniques are not exclusive to Southeast Asia, and in some instances it seems natural to use them if work is performed with no other than human power. I am not claiming that pounding and binding are the dominant or key elements in the entire technology, simply that they are emphasized in various ways and deserve our deliberate attention as possible "pulses" of the cultures in the region.

To show first the range of pounding and binding activities in a specific community, I draw on my field work in Eastern Indonesia (Adams ms., 1970, 1973, 1974). The two localities I studied in the eastern half of the island of Sumba happen to be very dry areas in which exploitation of plant life operates at a simple level. Material culture as a whole is limited. If remains were excavated by archaeologists, these sites would yield only simple pottery, postholes indicating rather large houses arranged in two rows within a cleared area, imported items such as bronze gongs, steel knives, and gold ornaments and, if preservation were good, a variety of

Figure 2. For a festival meal, women of East Sumba, Indonesia, pound rice competitively in special rhythms, often singing in chorus.

locally made thick cotton textiles decorated with beads and shells. Yet in the simple techniques the people employ, the accent on pounding and binding is evident, both in the daily routines and in more elaborated forms we call artistic such as gong orchestras and decorated textiles.

In East Sumba subsistence rests mainly on horticulture, both root and grain crops from dry fields, for which the soil must be prepared. Both men and women, working in groups, use the digging stick to prepare the soil. It is forced up and down in a pounding motion to break up the earth. For the few wet-rice fields owned by the royal families, men drive a herd of buffalo to plod over and over the watered fields, another kind of pounding. Food preparation, which is women's work, involves pounding to a marked degree. Rice, a festival food, is pounded with special ceremony to remove the husks (Fig. 2). Little food is eaten fresh. Usually crops are dried and prior to use are pounded (Fig. 3). Pounding does not facilitate storage of these foods (Yen ms.). Nevertheless, the favored ways of eat-

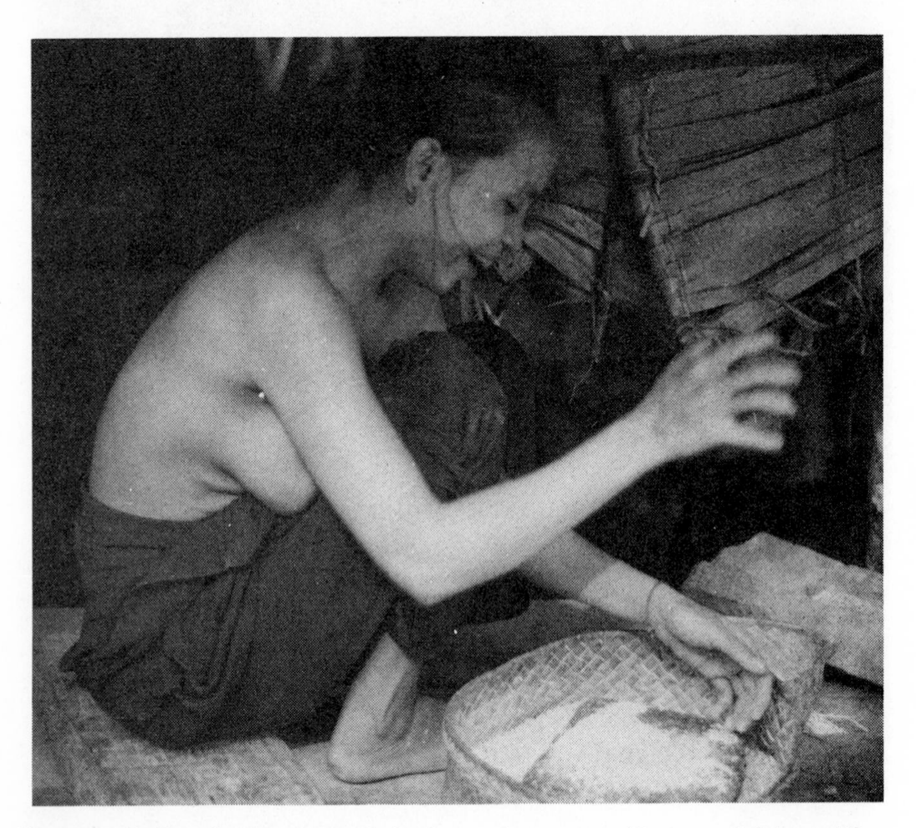

Figure 3. In East Sumba, the staple food, corn, is not ground but pounded, here with a stone on a stone mortar. Kapunduk district, E. Sumba.

ing corn, beans, seeds, peanuts, bananas, and other fruit is to dry and pound them into powder or semi-solid paste. The few special dishes prescribed by major rituals are pounded mixtures, for example, the sweetened vegetal patties at the year-purifying rite (*langa paraingu*) or the spiced banana-plant mixture *(lawar kukuta)* prized at weddings. In making containers, the women pound pots into shape by the paddle and anvil method.

Men, on the other hand, are proud of their skill in binding. They secure the wooden and bamboo house beams with bindings made of various plant fibers; for roofing, they separate dry grass into bunches and lash these in superimposed rows onto the light framework of the roof-rafters (Fig. 4). Men knot the nets for fishing which is a vital link in the seasonal food cycle (Fig. 5). The Sumbanese tend valuable herds of horses; binding devices such as bridles, tie-ropes, and lassos are all exclusively products of men's handiwork. For making houseposts or boats, men bind large trunks with heavy roots and cords and drag these in concert from the forests into the village. The proper activity for a man at home is sitting on his front veranda twisting cordage on his thigh.

I present these tasks in terms of men's and women's work because the Sumbanese stress these associations, not because I am proposing a consistent association throughout Southeast Asia of any one process with one sex. It is clear (Adams 1971) that in certain activities the roles are reversed. In smithing men pound the iron (although today because of imported finished products and travelling regional specialists there are few local smiths). In textile work women are intimately concerned with binding both as wrapping and as tying.

On Sumba, as in Southeast Asia generally, all the work associated with the making of cotton textile yarn for cloth belongs to the sphere of women's work. Here pounding is an initial keynote. The sound of beating raw cotton rings through the small valleys in the dry season. Two actions of pounding are repeated throughout weaving: one, the beating-in of each transverse yarn (weft) with a heavy wooden sword-like stick and, two, the pounding of the surface web (to help position the yarns) with another heavy flat stick (*rada*) (Fig. 6). In conversation about weaving, people will explicitly mention these sounds and mimic the beating action as characteristic of weaving. In legends, the hero's encounter with a young woman is announced by the phrase: "He heard her at her weaving."

If we turn to what we consider "the arts" that the Sumbanese have developed, we find their greatest interest focuses on gong orchestras made up of several men pounding on bronze gongs of various sizes. These sounds, essential at sacred rites, enliven every important occasion and demand the highest energy output. At small rustic events, sounds are created by stones that are pounded on other stones placed in the ground.

Figure 4. Lashing bundles of tough dry grasses onto the framework of a house roof. This work must be accompanied by a prayer-sacrifice rite and a festival meal for all participants. Kapunduk district, E. Sumba.

Figure 5. In the dry season men gather on verandas of local leaders' houses to mend their nets, twist cordage, make ropes for their horses and to tell stories. Here a prince of Kapunduk district, E. Sumba, with his personal attendant.

Ritual dancing for men stresses stamping and, for high admiration, simultaneous beating of a large drum (Fig. 7). The largest festival of life (in contrast to funerals), which is a high point of organization and artistic efforts, occurs at the dragging of massive stones to be set up as monuments to deceased men by their successful descendants (Fig. 8). Here binding is the major, explicit concern throughout the day or days as the men, to the accompaniment of chants recounting the deeds of the ancestors, drag the stones from distant hills to the village with bindings of cord-like roots. (In legends concerning this custom the hero's success in dragging hinges on the magical powers of the binding he uses).

Figure 6. Weaver Ana Hida seated at a backstrap loom. Her left hand touches the sword-shaped beater, now standing upright to open a passage for the transverse yarns (weft). The sword-beater is struck against each inserted weft to secure it in place. She will also pound the warp surface repeatedly, to help position the yarn, with the light colored flat stick (*rada*) lying above the round bars (shedroll and heddle). Kambera district, E. Sumba.

Figure 7. At a dry-season festival, men engage in dancing displays in which stamping steps are prominent. Simultaneous beating of large drum at the dancer's left is much admired. At right men beat old imported gongs. Kapunduk district, E. Sumba.

Another art form which enjoys a prominent position in Sumbanese evaluations (Adams 1969) is colorful decoration of cotton garments, large flat pieces which serve as wrap-around costume for men and women on special occasions. For womens' dress, designs are achieved by special weaving techniques, chiefly by inserting and manipulating extra strands while weaving the basic cloth. A different decorative technique is employed for the men's cloths. Patterns are dyed onto the yarn strands before weaving. This method (tie-dye or resist-dyeing) specifically employs binding to resist the dye. The salience of binding is acknowledged in the terms for this technique: in East Sumbanese, *hondu,* to bind, and in Indonesian language, *ikat,* meaning binding.

A general description of this technique and the illustrations in Figures 9 and 10 are sufficient here to demonstrate that binding is the essential action of this method. (For details and step-by-step illustrations, see Adams 1972). On Sumba, women practice ikatting on the lengthwise (warp) yarn strands only, as these will dominate the surface of the woven cloth. In ikatting, the women form the design by tying palm-leaf strips tightly around small bundles of warp strands (prior to weaving) in order to cover an area that conforms to the desired design. The bound hanks of yarn are dipped into dye solution which does not penetrate the tightly bound areas. After the bindings are cut away, the undyed portions appear as light-colored shapes against a darker, dyed ground.

Figure 8. In honor of the grandfather of Hina Malutaka of Mara Wua, Kapunduk district, E. Sumba, Hina sponsored a stone-dragging feast. The large stone, wrapped in grasses at lower left, is pulled by long, thick cord-like roots. For prestige the host invites many more men than are needed to help. Dragging the stone a short distance took one full day because of the irregular route taken and the frequent breaking of the bindings. In myths, the hero may successfully drag a stone with a single strand of magic fiber.

It is important, especially for those tempted to compare binding and pounding in Southeast Asia with other ethnographic areas, to note the sphere in which these techniques are prominent, that is, in major communal enterprises: preparing the soil, modifying basic foodstuffs, making costume, building houses, tending horses, dragging logs and stones, dancing, and music-making at festivals. All are central efforts of the Sumbanese community.

In Southeast Asia as a region, pounding and binding processes are significant in relation to artistic products. Several kinds of art for which the region is known involve the pounding process: embossed silver bowls and boxes, laminated swords and daggers, and orchestras in which the instruments of the gong family predominate. In Java, these musical ensembles are called *gamelan* from *gamel,* hammer. (For discussion of these arts, see Wagner 1959; Draeger 1972; Sheppard 1973; Maceda 1974). Yarn-fiber work executed by women is ubiquitous. It can be found in every ethnic group (except small nomadic gatherers), and a widely favored type of decoration is achieved by the binding method, *ikat.* Further, people at all social levels may appropriately practice these activities. We find reports

Figure 9. A preparatory binding step in the *ikat* technique. The craftswomen are combining layers of warp yarn, which has been separated into sets by bindings. The women are concentrating so that right sets will be matched and then bound together. Craftswomen: Hana Ata Endi and Rambu Windi, Kambera district, E. Sumba.

of princes and chiefs who were proud of their smithing skills. It is interesting to recall Raffles' observations (1830:95, 187) in early 19th century Java of a loom in every household, high and low, and of the pride expressed by husbands, both elite and commoner, in their wives' textile handiwork.

In many communities throughout Southeast Asia the work involved in preparing yarn and in weaving a decorated cloth represents the effort that is technologically the most complex. Producing cloth requires skillful deployment of a multitude of strands of cotton, hemp, or silk yarn. Critical to the successful ordering of these potentially straying strands are forms of bindings that secure them in fixed relation to each other (see Adams and Brandford 1976). The initial preparatory step of creating the warp surface (web) requires wrapping yarn around separated and parallel bars; cords interlaced transversely in this web maintain the separation of the upper and lower layers of warp that is essential for later weaving

Figure 10. The many layers of warp have been bound to a frame. Using palmleaf strips, the craftswoman is binding small groups of yarn strands in the shape of the desired patterns prior to weaving. Much of this warp has already been bound. These tightly bound areas will prevent dye from penetrating. This resist-dyeing technique produces designs by a light-and-dark contrast of colors. Kambera district, E. Sumba.

(Fig. 11). Weaving itself, that is, interlacing the wefts into the warp web, is essentially a form of binding. Wefts bind the warp together.

In Southeast Asia, elaborations of basic weave are generally obtained by means of supplementary weft techniques, that is, extra strands are inserted and manipulated as wefts; thus they are special forms of binding the warp (Fig. 12a and b). (Supplementary warp weaving occurs in small proportion to weft work). To see this preference as significant we might compare Southeast Asia briefly with another intensive weaving area, ancient Peru. Although supplementary techniques are well exploited in the Andes, there is also a great variety of alterations of basic weave, not connected with binding acts, such as spacing of yarns, ommission of wefts, differences in yarn thickness, and frequency relationships. In Southeast Asia we find mainly simple weaves with decoration coming from supplementary weft, a form of binding.

Ikatting, the other major form of textile decoration, is, as we have seen in the warp-ikats of East Sumba, a complex exercise in binding. In West-

Figure 11. Two women wrapping yarn around parallel bars (separated by half the length of the desired cloth) in order to form a warp. Note the transverse bindings which separate the warp into upper and lower layers essential for later weaving. In order to produce two costume cloths for a man's outfit, eight sections like this are needed. Craftswomen: Ana Hida and Hana Ata Endi, Kambera district, E. Sumba.

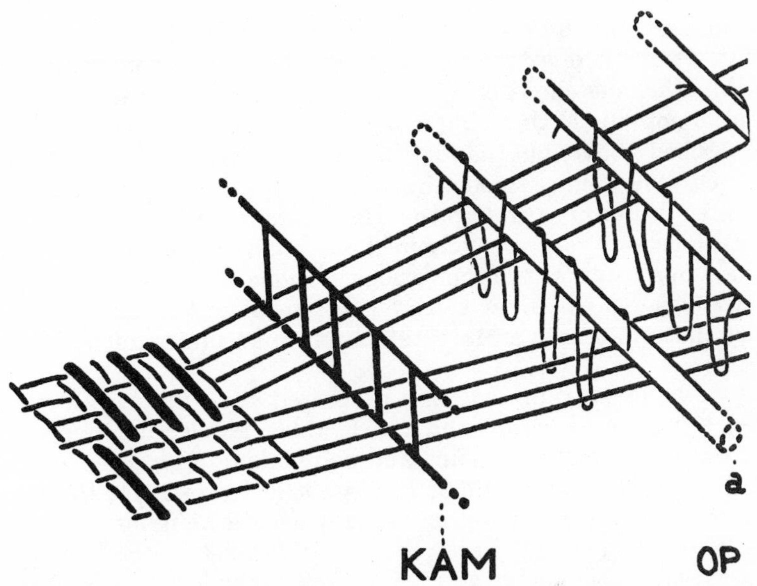

Figure 12a. Diagram of supplementary weft weave (in heavy black lines between the structural wefts) on a loom. Interlaced wefts bind the warps together.

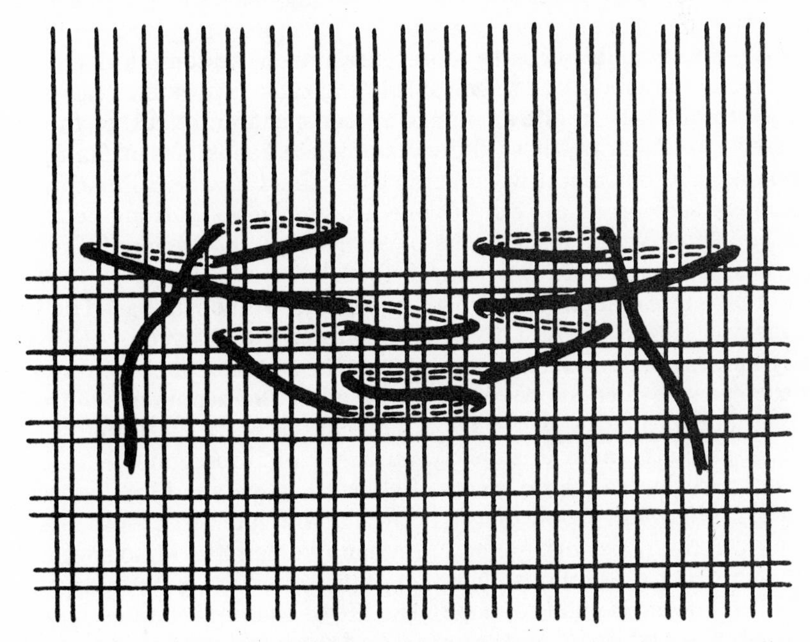

Figure 12b. Diagram of supplementary weft inserted during weaving with a needle or by hand. (Diagrams from Jager Gerlings 1952).

ern Indonesia and on the mainland plains, ikat binding is applied to the weft yarns and in one community on Bali, both warp and weft are bound, dyed and then matched during weaving. In all three variations, binding is essential not only in the tying-in of the design by means of leaf bindings, as described above, but also in the various stages of preparation of the yarn. On the tying frame, the bundles of yarn to be tied must be separated uniformly through the use of interspersed crossbindings, and several layers of yarn are combined and bound together in certain ways so that upon being woven the various sections of the dyed pattern will fit and the designs will show clearly (Fig. 9). Even this general summary shows that binding is an integral action in an important technical process.

As we have seen, producing a decorated cloth involves a sustained, varied sequence of action and considerable intellectual finesse. It demands a great amount of time and painstaking effort (see Adams and Brandford 1976; Adams 1969; Pelras 1962; Tietze 1941; Bühler 1972). Decorated cloth is important in its many uses as hangings, banners, costume, and gift exchange. Particularly in Indonesia, Dutch ethnographers realized that textiles carry a heavy load of symbolic meanings (for review see Jager Gerlings 1952; Solyom 1972). Textile hangings placed around a ritual area (inside or outside a structure) may serve a spiritually protective function; cloth banners on wooden poles may signify a combination of male and female elements representative of the community; costume may make statements as to social role and rank. Symbolism has emerged most prominently in gift exchange systems characteristic of Indonesian societies in which textiles, defined as female goods, are exchanged with weapons or other metal work explicitly valued as masculine things, the exchange serving as a form of contract on the obligations between the parties. (For recent discussions of symbolic importance of Indonesian textiles, see Bruner 1974; Gittinger 1972, 1974).

Perhaps less evident is an abstract kind of symbolism based on yarn as connective tissue. Certain highland rituals employ yarn or cloth symbolically and explicitly as a bridge between the human sphere and the world of the spirits which shows, in my view, that the people not only have a skilled and intimate awareness of the character of these materials but also appreciate them abstractly as a set of interconnected elements. Specific examples from recent fieldwork (Lyman 1968) among the Hmong (Miao) hill people of northern Thailand illustrate these uses.

In a ceremony performed for a pregnant woman, the Miao specialist using shamanistic techniques conducts a ritual which aims to capture the spirit of a dying person to ensure the life of the forthcoming child. First he stretches a white hemp thread around the interior house beams, lays a shuttle on the altar (during weaving a shuttle passes back and forth among the warp yarns making the essential connections between warps and wefts) and covering his head with a cloth begins to enter a trance-like state. When the shuttle falls to the floor, it serves as a bridge for the spirits to mount him. Before he performs the final critical rite, a thread is tied from the shuttle to the symbolic tree over his head and thence to the door, as a "path for the incoming spirit."

When a Miao couple wishes to have a child they may stretch an embroidered cloth from the door to their sleeping compartment as a "bridge" for the spirit to enter. In these phases of ritual, we see that the connective character of spun yarn, of weaving actions, and of woven cloth is being used as symbolic means to fulfill the purposes of the rites.

To turn now to the specific emphasis on binding rituals, I would first mention Hans Kauffmann's survey (1960, 1968) of the widespread occurrence of the thread cross (or god's eye), formed by binding yarn around crossed sticks, as an integral feature of rituals in Southeast Asia, thereby suggesting the importance in ritual of binding in itself.

On the mainland one of the most commonly reported binding acts in ritual is the binding of yarn around the wrist of a departing visitor as a guarantee of good welfare. Binding the wrists of the bride and groom is part of the traditional marriage ceremony in the lowlands (Fig. 13). Within the context of local belief, ritual binding acts function also in relation to serious concerns as in the following specific example from a lowland Thai village near Bangkok (Hanks 1963:66). After the birth of an infant, sacred cord is put with other symbolic things on a winnowing tray with the newborn, and all are tossed up together and then given to the mother. Thread is tied around the wrists and ankles of the infant in order, according to the explicit binding invocation, to seize and secure (to "root") the soul to the infant's body. This rite is seen as granting the infant membership in the human family. According to Hanks' further discussion of Thai beliefs, the fear of loss of the infant's soul is more than sentiment. A child is an important parental investment with an anticipated return upon its maturity. In return for the trouble of bearing and raising them, children are expected to work for the family until marriage. Further, a boy may earn merit for his parents by becoming ordained as a monk; a girl should care for her parents in their old age. If the child moved away, however, the parents lose out on their anticipated reward. Waywardness of spirit thus constitutes a continual threat, and parents employ "magical" or symbolic means to ensure that the child should stay close by. This symbolism is not an ornamental intellectual gesture but aims at the achievement of vitally important ends.

It would be neither feasible nor appropriate here to attempt to show the semantic levels at which binding homologies can be found in conceptual orders. I wish simply to point out an observable similarity between

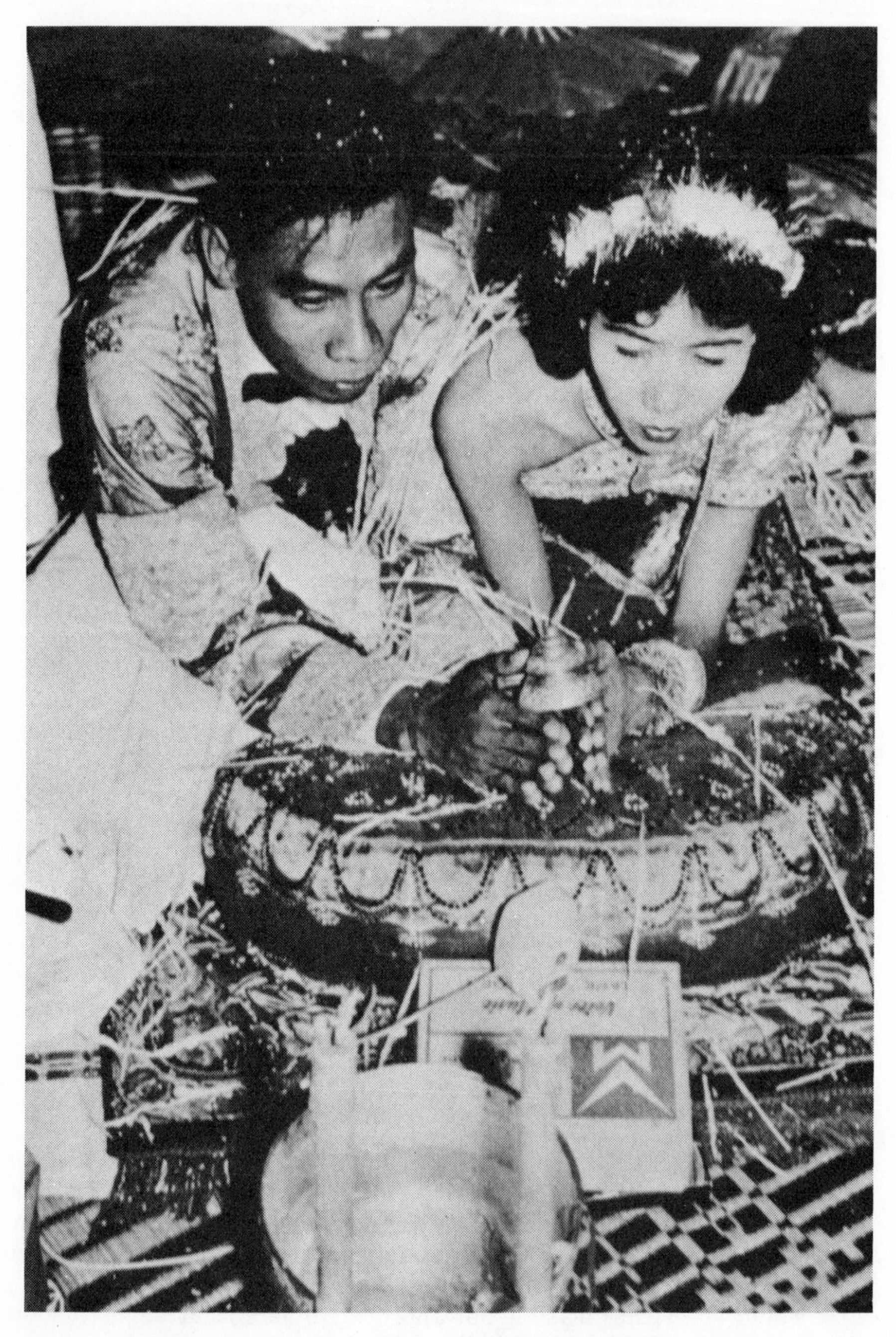

Figure 13. Bound wrists mark a traditional marriage ceremony in modern Cambodia. (Photo: Tooze 1962).

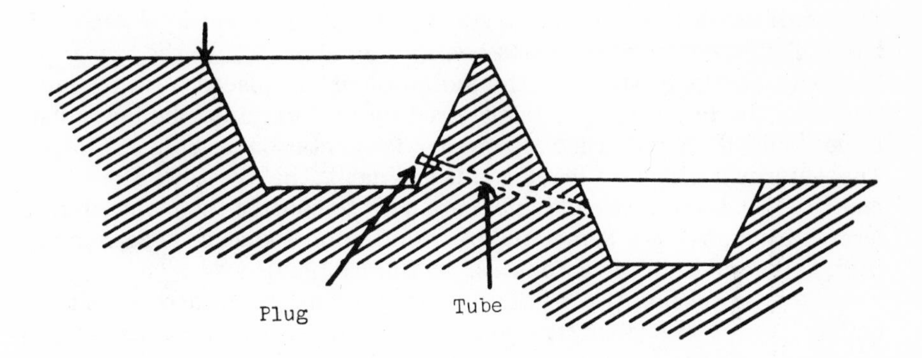

Figure 14. Pits used in preparation of indigo dye by the Yao, an ethnic group occupying several provinces of South China and northern Southeast Asia. (Diagram from Wist 1938).

technical and ritual practices which recurs with sufficient constancy to suggest a preference for a certain manner of behaving that in turn elicits the term "style." The comparisons above are in the form of a simple parallel: acts of binding in cloth-making and acts of binding in ceremonial contexts.

One could point to more complicated similarities, however, such as those that occur in Southeast Asia between a favored style of processing vegetal matter and of ordering funeral ritual. Both activities are characterized by emphasis on decomposition stages in order to obtain a finer essence which is the desired end product. For example, the staged processing of vegetal matter often develops in a three-step sequence—decomposition,[1] transfer, and further disintegration—the same sequence in which the royal funeral ritual of the mainland unfolds. Difference in scale between acts of vegetal processing and the great royal funeral may seem so large as to daunt the attempt to compare them, but under all the elaborations of royal abundance and display there are three well-defined stages in royal death ritual that clearly emerge. The examples of vegetal processing techniques that I supply here concern textile dyes, foods, and habitual stimulants.

The first illustration of this three-stage sequence is provided by the processes used in the manufacture of blue-black dye, the most characteristic and widespread color for decorating Southeast Asian cloth. One of the simple versions of making blue dye from indigo can be found among the Yao who live in the highlands stretching across the eastern provinces of South China (Kweichow, Kwangsi, and Kwangtung) and the northern areas of Southeast Asia (North Vietnam, Laos, and

Thailand) (Wist 1938). Two pits are dug side by side, but at slightly different elevations with respect to one another (Fig. 14). In the higher pit the roots and leafy stems of the indigo plant are placed in water until they rot; the liquid is then transferred by draining into the lower vat. Lime is added, and after a week or so of fermentation[2] a sediment forms. The water is drained off leaving this sediment, the dyestuff. More commonly in Southeast Asia clay pots are used to contain the plants during fermentation, but the same kind of transferring sequence is carried out (Fig. 15) (For details of procedures, see Bühler 1939).

In food processing, I consider fermentation a form of decomposition; boiling is a milder form in that boiling breaks down fibers and makes food more accessible to further decomposition. The useful aspect of fermentation is preservation through longer storage capacity (Yen ms.). Vegetal matter can also be toxic, unpleasant in taste, or indigestible and is often treated by various methods such as drying (in humid areas not always feasible), smoking, and fermentation to alter these properties. In Southeast Asia, food may be steamed or boiled (decomposition), made into a paste, buried in a container with hot ashes (transferred), and fermented

Figure 15. Two sets of pots used in indigo dye preparation in the Kambera district of East Sumba. At left, pot contains rotting indigo plants; at right the second stage of dye preparation: the pot contains liquid drained from the indigo plants and mixed with lime. This mixture will form a sediment that is the dyestuff. Basketry covers lie beside pots.

(further disintegration). An example from the Semang people in the Malayan rainforest of the mainland (Pelzer 1945) illustrates the three stages involved in food preparation through fermentation. Part of the year, the Semang live on initially poisonous seeds which have been boiled (decomposition), pounded and packed in barkcloth or bamboo tubes (transfer), and buried for a month until fermented (further disintegration) and edible.

Generally, however, fermentation is developed in underground pits or in tightly covered containers (bowls, pots, pits, and vats are important elements in the processing of these vegetal substances). The range of foods treated by these methods shows that the practical problems mentioned above (e.g. toxicity, indigestibility) are not always at issue. Lebar (1967) has pointed out that a large variety of fish, vegetables, and shellfish is preserved in Southeast Asia (and Oceania) through fermentation either because the food spoils or because the people like the astringent flavor.

These techniques were probably much more common prior to the 19th century when rice became more widespread as a staple in mainland and insular Southeast Asia (Delvert 1961). In the early 19th century, for example, the Javanese diet consisted mainly of root crops, fruits, leaves of plants, wild roots, and tubers (Pelzer 1945). The decomposition processes described above were not limited to the poor but were familiar to all classes. The great labor involved in these processes testifies to people's strong willingness to go through all these stages in order to produce a transformed finer essence which may be lesser in quantity but which has a higher culturally rated value than ordinary foodstuff.

These food and dye products are not only characteristic of Southeast Asia from the outsider's viewpoint but the people themselves accord such processed stuffs special value. A glance at any collection of Southeast Asian costume will quickly convince one of the prominent role played by blue dyes in the decorative enhancement of garments. Among foods, in Thailand, for example, boiled rice provides the staple dish, but the Thai do not consider a meal properly complete without the traditional decomposed or fermented, spicy fish-pastes. Palm wines are widely enjoyed. Preparation of the special fermented drinks shows a three-stage sequence, as for example in the following Thai recipe for rice-liquor (Bowring 1857): the rice is first steamed (decomposition), sprinkled with ginger and spices, wrapped in plantain leaves (transfer), and in 24 hours a sweet vinous liquor exudes (fermentation as further disintegration). If kept longer, the drink becomes intoxicating. Distilling (another processing technique for finer essence by fermentation and transfer) produces *arak*, a strong fragrant drink.

A distinctively Southeast Asian stimulant is preserved tea leaves, which are chewed. These are processed by fermentation in underground pits in the highlands of Thailand, Laos, and Viet Nam and distributed widely in the lowlands. The most widely used habitual narcotic consists of areca palm nut (betel) which may also be treated by pit-fermentation. It is interesting, too, that a more recent (probably 17th century) addition to Southeast Asian narcotic substances, tobacco, is also treated by fermentation. On Sumba and elsewhere, a chew of fermented tobacco is considered its most desirable form.

All these processes are humid transformations which frequently take a three-stage form: first, decomposition, then a transfer, and finally further disintegration leading to the production of a refined essence or remains which constitutes the desired substance of a higher order of value.

We know these traditions are of considerable historical depth beyond the period of ethnographic report, because ancient Chinese visitors to the southern peoples of China who share cultural traditions with many peoples of the Southeast Asia mainland, mentioned, for example, the drinking of wine from closed jars, the pickling of food, and the wearing of colorful clothing, especially streaked and spotted cloths, indicating the use of dyes (See Eberhard's reference to 10th century Chinese documents, 1968:273–4). The spotted cloths may refer to decorations achieved by reserve techniques best known today among the highland Miao (northern Southeast Asia and South China provinces) and in Indonesian batik, while the references to streaked cloth may be interpreted as a description of ikatting. Both these techniques are still practiced and use blue-dyeing (requiring decomposition as noted above) as the primary coloring process.

With this long tradition of preferential processing in mind, we turn to examine the sequences of a particular kind of ritual. I have selected the royal funeral rite in the lowland countries of Thailand, Cambodia, and Laos because it is the most elaborately developed form of ritual in Southeast Asia and, as such, can be taken as the fullest statement of a certain ritual style.

The procedures from the death of a king to the ultimate disposition of his remains will be outlined briefly to show that the main stages, marked by elaborate ritual, are decomposition, transfer, and further disintegration which produce the final remains that are accorded a higher value (Quaritch Wales 1931).

Upon completion of private rites held in the king's apartments, the corpse is fitted into a tall urn which is placed within another more elaborately decorated urn. This larger urn (4 to 5 feet high) is deposited on an altar in a great hall destined for this purpose, where it remains lying-in-state for several months (Fig. 16). The period during which the body decomposes is the first official stage of the funeral. The base of each urn is perforated so that the body may drain; the "sanies" are carried off each day with special rites. This stage is marked by major court ceremonies which require the presence of the royal family and members of the government.

The next ritual step is the great procession which transfers the urn containing the remains to the cremation ground (Fig. 17). To accent the idea of transport, the urn is borne by human porters on a great cart in the shape of a ship. In recent times, this processional stage has been marked by artillery, aviation maneuvers, and thousands of costumed attendants.

Figure 16. Lying-in-state of funeral urn in official hall of palace compound. Bangkok, Thailand. (Photo: Quaritch Wales 1931).

At the cremation ground, the urn is placed on top of a lavishly decorated stepped platform. Many persons pay their respects by adding something such as spices and perfumed woods to the pyre (Fig. 18). During the night, flames reduce the corpse in a further disintegration to charred remains. The next morning the finer essence is obtained. The ashes mixed with unguents are given the form of a human figure and turned to the east, west, then east again (the rising, setting, and rising of the sun in an explicit reference to a belief-schema of birth, death, and rebirth), wrapped in cloth, placed in a golden dish, and taken away in procession, eventually to be enshrined in a sacred compound (Fig. 19).

At the cremation site, the search for relics begins. King, Queen, nobles, and ladies take part seeking fragments of burnt bone some of which are brought in a gold plate to rest upon the palace altar; others are kept by different relatives. These remains are considered sacred and provide the

Figure 17. Royal funeral procession from palace compound to cremation ground. Bangkok, Thailand. (Photo: Quaritch Wales 1931).

Figure 18. At royal cremation ground, a large decorated pavilion supports the royal urn. During the day, people paying their respects add bits of fragrant wood. Actual burning of corpse takes place during the night. Bangkok, Thailand. (Photo: Quaritch Wales 1931).

Figure 19. After cremation, procession carries away remains which are revered as relics. Bangkok, Thailand. (Photo: Quaritch Wales 1931).

revered relics for which later religious structures and commemorative works are built. The result of the disintegration process is a finer essence.

The important point about the funeral as a whole is that death, instead of being presented as a mortal collapse, is considered by the sponsors and the public as the most important phase of a migration to a higher plane, from the status of king to one honored as a higher spiritual being (Quaritch Wales 1931). The ritual attention given to the decomposing body, the massive procession, the elaborate setting for the final reduction into significant remains illustrate the stages of decomposition, transfer, and further disintegration leading to the formation or reconstitution of a finer essence, stages which are similar to the sequence that was evident in the food, dye, and narcotic processing.

It may seem puzzling at the outset that human flesh, so like animal substance, should be processed in a manner conceptually comparable to vegetal (and fish) material. The funeral rites may be the ultimate fulfillment of the metaphor of man as plant which is, in my view, a widespread metaphor in Southeast Asian thought. (For demonstration of this metaphor in one region of Indonesia, see Fox 1971).

Simpler funeral ceremonies are characterized by similar stages. After preliminary burial, cremation or disposal by exposure to birds, the bones are collected, placed in an urn or other container, and conveyed to their final abode. This concern for what I call "processing the corpse into a finer essence" is widespread in Southeast Asia. As Hertz (1960) concluded after surveying Indonesian mortuary and secondary burial practices, the concern for the decomposing stage stems from a belief that the fate of the soul is related to the fate of the decomposing corpse. We may better understand at least the character and persistence of such important ritual procedures through an awareness of preferences for certain technical processing sequences.

Essentially what I am suggesting is that we look at the repetitive and structured techniques of processing materials in daily life as parallels to or possibly models for other activities, in particular, some of those that we call imaginative, projective, or more value-laden, such as ritual and art forms.

Although made in the context of a different cultural situation and different investigative purpose (that is, a search for explanatory models), Horton's observations (1964:99) convey this core notion:

> Now in technologically backward communities which have a relatively simple social organization and are not in a state of rapid and self-conscious change, peoples' activities in society present the most markedly ordered

> and regular areas of their experience, whereas their biological and inanimate environment is by and large less tidily predictable. Hence it is chiefly to human activities and relationships that such communities turn for the sources of their most important explanatory models.

However, a broader perspective might view both the technology and the ritual as problem-solving modes, the technical processes for handling changes in materials, while the ritual sequences or the aesthetic experiences are addressed to many kinds of problem situations that require other kinds of transformation. In this view the character of the procedures is, in some way, a choice made by the people, its constancy and use in several realms suggesting a stylistic expression. In the instances I have discussed, technological style seems to overlap with ritual style. Preferences in method manifest themselves similarly in both systems.

Why does one people choose a particular style over another if there are possible alternatives? It is tempting here to paraphrase an earlier suggestion I made (Adams 1972) in connection with a specific case study of the compositional principles of textile design within Sumbanese culture, that modes of problem-solving stem from deep-lying aesthetic preferences. Obviously that realm is difficult to investigate. As a practical matter. however, I hope this paper "raises consciousness" of new possibilities for research on both observable and ideological aspects of the relationships between technical process and other activities in society.

NOTES

[1]Decomposition is being used here in its standard meaning as the process of separation or resolution (of anything) into its constituent elements, dissolution of compound bodies, disintegration, organic decay, putrescence. (*Concise Oxford English Dictionary* 1971:664).

[2]In popular language the term fermentation conveys the notion of sensible effervescence. Modern science restricts the term to a class of chemical changes peculiar to organic compounds, produced by the stimulation of a "ferment." Various kinds of fermentation are distinguished by adjectives such as acetous, alcoholic, butyric, lactic, putrefactive, and so forth. Before the modern chemical definition, the term applied to all chemical changes exhibiting the character, superficially recognized, of effervescence or internal commotion in the substance and the resultant alteration of properties. (*Concise Oxford English Dictionary* 1971:985). I use the term in the latter sense, that is, as one general category depending on superficial recognition, because I am assuming that this most closely corresponds to observations of village people in Southeast Asia.

LITERATURE CITED

Adams, Marie Jeanne
1974 Symbols of the Organized Community in East Sumba, Indonesia. Bijdragen voor Taal-, Land-, en Volkenkunde 130 (2/3):324–347.
1973 Structural Aspects of a Village Art, East Sumba, Indonesia. American Anthropologist 75(1):265–279.
1972 Classic and Eccentric Elements in East Sumba Textiles. Needle and Bobbin Bulletin 55:1–40.
1971 Approach to Arts and Ceremony, East Sumba, Indonesia. Newsletter of the American Anthropological Association 12(3):5,20.
1970 Myths and Self-Image Among the Kapunduk People of Sumba. Indonesia (Ithaca) 10:80–106.
1969 System and Meaning in East Sumba Textile Design: A Study in Traditional Indonesian Art. Southeast Asia Studies, Cultural Report 16. New Haven, Conn.: Yale University Press.
Ms. Field Journals, Sumba, 1968–9.

Adams, M. J. and Joanne Brandford
1976 Review of 16 Craft Films on Highland Southeast Asia. American Anthropologist 78(3):720–722.

Bacon, Elizabeth
1946 A Preliminary Attempt to Determine Culture Areas in Asia. Southwestern Journal of Anthropology 2(2):117–132.

Bernatzik, Hugo
1947 Akha und Meau. Innsbruck: Wagner'sche Univ.-Buchdruckerei. (English Translation, 1970, New Haven: Human Relations Area File.)

Bowring, J.
1857 The Kingdom and People of Siam. Facsms. Reprint. New York: Oxford University Press, 1969.

Bruner, Edward M.
1974 Indonesian Homecoming: A Case Study In the Analysis of Ritual. Reading, Mass.: Addison-Wesley.

Bühler, Alfred
1972 Hanfverarbeitung und Batik bei den Meau in Nordthailand. Ethnologische Zeitschrift 1:61–81.
1939 Ikats. Ciba Review 44:1586–1611.

Concise Oxford English Dictionary
1971 Oxford: Oxford University Press.

Delvert, Jean
1961 Le Paysan Cambodgien. Hague: Mouton.

Draeger, Donn F.
1972 Weapons and Fighting Arts of the Indonesian Archipelago. Rutland, Vt.: Tuttle.

Eberhard, Wolfram
1968 The Local Cultures of South and East China. Leiden: Brill.

Fox, James J.
1971 Sister's Child as Plant, Metaphors in an Idiom of Consanguinity. *In* Rethinking Marriage and Kinship. Rodney Needham, ed. Pp. 219–251. London: Tavistock.

Gittinger, Mattiebelle S.
1974 South Sumatran Ship Cloths. Needle and Bobbin Bulletin 57(1/2):3–29.
1972 A Study of the Ship Cloths of South Sumatra: Their Design and Usage. Unpublished Ph.D. dissertation, New York, Columbia University.

Hanks, Jane
1963 Maternity and its Rituals in Bang Chan. Southeast Asia Program, Data Paper 51. Ithaca, N.Y.: Cornell University Press.

Heine-Geldern, Robert
1923 Südostasien. Illustrierte Völkerkunde. Georg Buschan, ed. II:801–934, Stuttgart: Stecker, Shroeder.

Hertz, Robert
1960 Death and the Right Hand. Translation of essays *In* Année Sociologique (1907–1909) Rodney and C. Needham, eds. Glencoe, Ill.: Free Press.

Horton, W. Robin
1964 Ritual Man in Africa. Africa 34(2):85–103.

Jager Gerlings, J. H.
1952 Sprekende Weefsels. Amsterdam: Royal Tropical Museum.

Kauffmann, Hans E.
1968 Das Fadenkreuz. Ethnologica, Part II 4:264–313.
1960 Das Fadenkreuz. Ethnologica, Part I (Cologne) 2:36–69.

Kunstadter, Peter, ed.
1967 Southeast Asian Tribes, Minorities and Nations. Princeton, N.J.: Princeton University Press.

LeBar, Frank M.
1967 Miang: Fermented Tea in North Thailand. Behavioral Science Notes 2:105–119.

Lechtman, Heather and Arthur Steinberg
In press The History of Technology: An Anthropological Point of View. *In* Proceedings of the International Symposium on the History and Philosophy of Technology, University of Illinois at Chicago Circle, 1973.

Loofs, H. H.
1964 Südostasiens Fundamente: Hochkulturen und Primitivstämme. Berlin: Safari-Verlag.

Lyman, Thomas A.
1968 Green Miao (Meo) Spirit Ceremonies. Ethnologica (Cologne) 4:1–29.

Maceda, José
1974 Music in Southeast Asia: Tradition, Nationalism, Innovation. Cultures 1 (3):75–94.

Mori, John L. and Jocelyn
1972 Revising Our Conceptions of Museum Research. Curator 15 (3):189–199.

Oswalt, Wendell H.
1973 Habitat and Technology, The Evolution of Hunting. New York: Holt, Rinehart and Winston.

Pelras, Christian
1962 Tissages Balinais. Objets et Mondes 2(4):215–240.

Pelzer, Karl J.
1945 Pioneer Settlement in the Asiatic Tropics. New York: American Geographical Society.

Quaritch Wales, Horace
1931 Siamese State Ceremonies. London: Quaritch.

Raffles, T. S.
1830 The History of Java. 2nd Ed. London: J. Murray.

Richardson, Miles, ed.
1974 The Human Mirror. Baton Rouge: Louisiana State University Press.

Sheppard, Mubin
1973 Taman Indera, Malay Decorative Arts and Pastimes. New York: Oxford University Press.

Solyom, Garrett and Bronwen
1972 Textiles of the Indonesian Archipelago. Honolulu: University of Hawaii Press.

Spier, Robert F.
1970 From the Hand of Man: Primitive and Preindustrial Technologies. Boston: Houghton-Mifflin.

Tietze, Kathe
1941 Sitten and Gebrauche beim Söen, Ernten, Spinnen, Ikatten, Färben und Weben der Baumwolle im Sikka-Gebiet (östliches Mittel-Flores). Ethnologica (Leipzig) 5:1–64.

Tooze, Ruth
1962 Cambodia: Land of Contrasts. New York: Viking.

Wagner, Fritz A.
1959 Indonesia, the Art of An Island Group. New York: McGraw-Hill.

Wist, Hans
1938 Die Yao in Südchina. Baessler Archiv 21(3):73–135.

Yen, D. E.
Ms. Indigenous Food-Processing in Oceania. Paper for the 9th International Conference on Anthropological and Ethnological Sciences, Chicago, 1973.

3

Technology and Culture: Technological Styles in the Bronzes of Shang China, Phrygia and Urnfield Central Europe

ARTHUR STEINBERG

Massachusetts Institute of Technology, Cambridge, Massachusetts

INTRODUCTION

In the cultural manifestations of a technology a style is discernible that is a specific statement of how to use and work materials. This technological style can be characterized and compared to other styles independent of the art style, the symbolization, or the subject that is being rendered by a particular technique.[1]

Technology as an integral component or subsystem of culture interacts in a number of complicated but definable ways with other subsystems of culture. In describing a technology we are concerned not only with the specific techniques involved in making things, but also with the nature of raw materials procurement (trade, recycling, markets, etc.), with the ancillary technologies required to support the technology in question (e.g., if bronze casting is to be studied then one must also consider ceramics, mold making, refractories, furnace design, and fuel procurement), with the particular choice of material for the object, and with the choice of processes for working that material. We must also concern ourselves with the relationship between the function of the technological product and its appearance (which means paying attention to the quality and care of design and production as a function of use). Finally we are concerned about the deposition or findspot of the object, since we are here generally concerned with archaeological remains. One of the most important things to consider is the socio-economic status of the craftsman in the society as well as his relation to patron or market. In short, we are considering the technology, its practitioners and products, as intimate parts of a larger whole. When we come to speak of technological style we are essentially looking at how that "larger whole"—namely the culture—seen through its products, has interacted with various features of the technology to give it a specific form.

Our study of technological styles is complicated by the fact that we, the modern observers, are far more sensitive to the pictorial or iconographic components of a style than to its technical ones. We have studied ancient and medieval art so extensively that we can distinguish extremely fine nuances from hand to hand and school to school. This is not true for aspects of technique, technical skill, manner of treating a particular material, choice of material, and so on. We do generally realize that there are limited ways of treating a material such as bronze, for example—it can be hammered (raised or forged), cast, engraved, traced, inlaid, patinated, and so forth. But we fail to consider that the craftsman in a particular culture who wishes to make an object for a specific function has as his limiting conditions not only the general physical properties of bronze, but also one or relatively few ways of actually working and finishing the material which are dictated by his own knowledge and experience. Both are clearly functions of his cultural ambience. Thus either hammering or casting is generally the "accepted" way of making something of bronze, but once that general manner of working has been determined a host of other related choices such as the nature of the alloy, the method of detailing, of finishing, of coloring, also must be made. It is the particular constellation of these culturally known and acceptable ways of working materials for specific ends that constitutes a technological style. And though we have paid too little attention to technological styles in the past and, as a result of that negligence, cannot yet distinguish them as clearly as we can pictorial or art styles, I think that ultimately we will find these technological differences quite as important and informative as the others.

Before going on to actual examples of style in technique, I should point out two areas of confusion. First is the phenomenon "skeuomorphism" or imitating one material and technique by another (Spier 1970:109). Thus the 'Ubaid representations of copper and stone axe-heads in clay, the Sialk representations of metal container shapes in pottery, the common representation of Neolithic basket and skin container forms in pottery, are all examples of imitations which are not easily comprehensible to us, though we might consider, as a parallel, our own culture's translation of metal and ceramic forms into plastic. The modern example illustrates cheaper, more easily mass-produced techniques substituted for costlier, slower means of production. The earlier examples may be motivated by a similar economic drive, which would certainly tell us something about cultural values, but we cannot be sure, for aesthetic, ideological, or religious reasons may also underlie these earlier translations. I raise this issue because it will influence our perception of a technological style if we find that a main characteristic of a certain culture's technology is making one material look like another, hiding as it were, the basic physical properties of a material and process by making them look like different materials, produced by different means (it is, after all, not reasonable to think of casting and riveting clay axe-heads or weaving pottery containers). The translations taking place in these instances, in and of themselves, have important stylistic meanings and are undoubtedly significant indicators of a culture's ideological and aesthetic content.

The second point of confusion arises from the fact that much of what we study in terms of styles is blurred by diffusion. Techniques, like iconographic or symbolic systems, do not generally develop in cultural isolation but are spread from place to place by itinerant craftsmen, traders, and travellers, and are then superimposed or added to indigenous motifs and techniques. It is often difficult to unscramble the new mix into its original constituent parts of local materials, ways of working them, themes, and motifs, and to analyze what has happened to them when the new stimuli have arrived. To a large extent this process of diffusion leads to homogenization, as is evident in the present spread of North American and North European industrialized norms and forms to the rest of the world. Fortunately for those of us interested in studying some of these processes at earlier times and in different places this tendency towards homogenization occurred much more slowly and was less complete, so that we can still detect specific technological styles operating at certain times and watch them being replaced or altered by new ones (so it is with the introduction of lost-wax casting into China in Late Chou or Spring and Autumn times to supplement sectional-mold casting techniques; or the replacement of local Italic ceramic styles by the introduction and imitation first of the Greek Corinthian style and then of the Attic Black and Red Figure painting; or the spread of Roman sigillata ceramics throughout the early Empire replacing many of the local wares.)

Though one can cite many clear examples of technological styles exhibited by ways of working materials that are specific to a particular culture, such as Black and Red Figure pottery decoration in Archaic and Classical Athens, Etruscan gold granulation, Sung celadon glazing, Japanese steel samurai sword-smithing, or Andean "depletion gilding" (Lechtman 1973), we will focus here on materials and ways of treating them that allow more conclusive cross-cultural comparisons. The discussion will deal with bronze ritual vessels that are probably all connected with wine serving and are all found in burials in three different cultural settings, illustrating dramatically three distinctive technological styles of bronze-working: that of Shang China in the period after 1500 B.C.; of Phrygia, in Anatolia, around 800–700 B.C.; and of Central Europe late in the Bronze Age, around the 12th century B.C. Before looking specifically at each bronze-working technique, it will be helpful to sketch briefly some of the major features of the culture of which it is a product. We are, unfortunately, sorely hampered in this endeavor by the very limited state of knowledge of these archaeologically known cultures. The approach developed here would be considerably enhanced if technologies in ethnographically or historically documented cultures were investigated so that we might arrive at a better sense of the nature of the fit between technological and other cultural features.

SHANG CHINA

The first technological style to be examined is the casting of ritual bronze vessels in Shang China. The long-lived Shang dynasty (ca. 1600–1027 B.C.) marks the emergence in China of nucleated, Bronze Age, urban centers with marked social stratification, an aristocratic political organization, high-status crafts, writing (evident in the oracle bones), warfare, and so forth (Chang 1968, 1974; Wheatley 1971; Treistman 1972). The evolution appears to be similar to that in Mesopotamia and other pristine societies, where this kind of nucleation, growth, and stratification develops from more egalitarian farming-herding villages with less elaborate technologies except for pottery making; this latter phase is generally designated as Neolithic (Service 1975). The Chinese evidence at sites like Cheng-chou and Anyang reveals a centralized urban cluster with a walled precinct containing temples and "lineage houses" on raised earth platforms, with dispersed outlying agricultural peasant villages of subterranean houses, and royal cemeteries. The high-status crafts including bronze casting, fine pottery making, jade working, and bone and stone carving, are often located near the civic center and seem to be associated with one or another "lineage house" (Chang 1968:199–252, especially Figs. 72, 85, 87). The craftsmen's houses are of the earth platform type rather than semi-subterranean, which is regarded as an indication of the high status of the craftsmen and their crafts. Molds for bronze ritual vessels are found in these combined craftsmen's houses and workshops, while the vessels themselves are invariably found in the extraordinarily rich royal burials outside the city, along with jades, carved bones, bronze weapons, and human and animal sacrifices. Inscriptions on the vessels themselves tell us that they were used for food and drink in the all-important cult of ancestor worship, which consisted of propitiating the gods and ancestors in the next world in return for a successful life in this world. The society was dominated by two dynastic houses that traded off political control in a cyclical fashion, each in turn keeping its relatives in positions of power. This seems to be a form of highly stratified chiefdom well on its way to becoming a kingdom, since the chief or king appointed either relatives or nobles whom he elevated to positions of power to help him rule. The crafts were an integral and supporting part of this power structure, since their products were required to keep the world of gods and men in balance with one another. As a result, the craftsmen, who appear to be kin-organized and probably associated with specific lineages, were elevated to unusually high status positions. Commensurately their crafts are of an extraordinarily high quality.

This is not the place to enter the debate over the origins of the Chinese bronze-casting industry, since we are essentially interested in its function and style rather than its origins (Chêng 1974; Gettens 1969; Barnard 1961; Smith 1972). But it is worth observing that the bronze-casting industry which culminates in such productions as the vessels from Anyang or Anhuei does not show a clear development out of a simpler casting tradition. Its antecedents are an occasional spear, knife, hook, or bell cast in simple, cored, two-piece molds at sites such as Ehrlitou. But the ritual vessels of the Neolithic village are pottery vessels, or put another way, the ancestors in shape and design of the bronze ritual vessels are found in ceramic examples from Neolithic burials (whether these are status-differentiated burials is not clear) (Loehr 1968; Watson 1962).

Figure 1. Late Shang *Fang I* (Fogg Art Museum No. 1943.52.109), probably from Anyang. Lid and body were cast separately, but the mold for each was built up in separate previously worked sections by incising and impressing various parts of the design into each section. It appears that separate slightly different patterns were used for each panel, and the fine incision was then added at the end as a space-filler. The vessel with lid is about 30 cm. high.

This connection of the bronzes with ceramics is important, for even the earliest known examples of these bronze vessels are made, in the characteristically Chinese fashion, with sectional ceramic molds. Description of a Late Shang example in the Fogg Art Museum (Catalog No. 1943.52.109) will illustrate this more clearly (Fig. 1). A model of the desired vessel was made in some material; modeled over this were sections of ceramic mold with parting lines along the midline of the raised flanges. Designs were impressed and carved into the mold, in the negative, in several different stages (Fig. 2). The mask of the *t'ao-t'ieh,* its ears, horns, and eyes were all impressed separately into the mold from what appear to be previously carved and incised models of wood or possibly clay, all slightly different in size; the dragons and other monsters above and below the central masks were impressed in the same manner. Then the *lei-wen* meander pattern was added freehand as a design filler

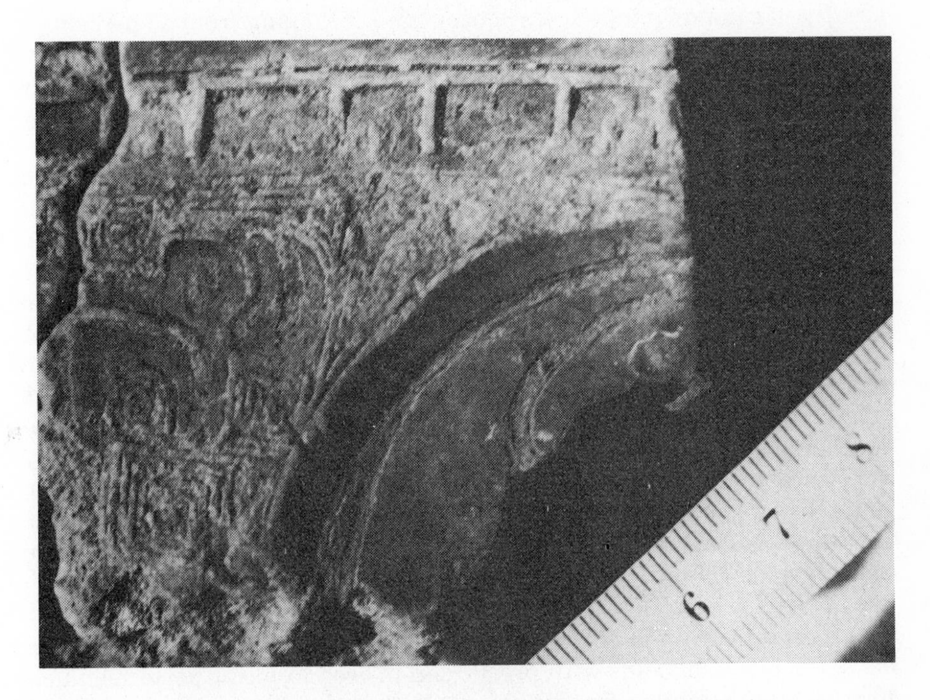

Figure 2. Late Shang clay mold fragment for a bronze casting, probably from Anyang (Sumitomo collection). The indented section in the lower right of the fragment was impressed into the mold while the clay was leather-hard, the whirligig pattern was scratched on it, then the *lei-wen* pattern was incised around it. The scallopped border at the top was probably stamped into the mold. (Photo: courtesy of C.S. Smith).

by incision directly into the smooth surface of the mold. This procedure was followed on all four sides of the vessel body and lid. Some additional methods of building up patterns are found on other vessels including the insertion of whole preformed clay slabs, decorated in the negative, into the preshaped mold sections, and of animal heads that are modeled in three-dimensions and could not possibly have drawn from the mold (Smith 1972:107–114). Both of these techniques still require closer study before we can be certain how they were executed. The basic technical procedures, as described for the Fogg *Fang I,* consist of carving, impressing, and incising the design either into various separate patterns or directly into the mold. After this the mold was disassembled, the model was cut down by an amount equal to the desired wall thickness of the bronze, or a new core was made; the whole mold (mortised and tenoned together) was then reassembled over the core with appropriate spacers (chaplets) centering the core. The mold was then baked and the bronze cast into it while the mold was still hot. The metal used for bronzes like the *Fang I* varies greatly in composition, the tin concentration ranging from 13 percent to 17 percent and lead often present in the 1 percent to 5 percent range (Gettens 1969: 33–56). After removal of the mold the bronze was finished by tooling and surface abrasion, both local and general, and possibly patinated.

Several characteristics of this technology are noteworthy. First and foremost, the whole process relies essentially on the carving and other manipulation of a ceramic mold. Bronze casting is always dependent on mold making, but in the lost-wax or sand-casting methods common in the West the stress is on carefully working the model on which the mold is to be made and later tooling the finished casting to bring out details. Not so in Shang China where the main work was done on the interior surfaces of the mold from which the bronze was to take its impression, and there was relatively less tooling of the finished casting, notwithstanding the extraordinary intricacy of much of the design. This design came out extremely well for a number of reasons, including: painstaking care in preparation of the molds; casting metal into such molds while they were very hot; using alloys that flowed well and chilling them slowly; and generally observing good foundry practice such as proper gating and risering which are obvious prerequisites for these superlative castings.

Important elements of this technology consisted in impressing design elements into the molds with precut and incised forms or patterns much like the stamp-impressing of pottery. On the other hand, the carving and incising aspect also has clear counterparts and antecedents in worked materials like wood, bone, and stone all of which are found in Shang and earlier contexts. The assembling of intricately worked, preformed sections of molds may be a kind of reflection of intricate woodworking as in the making of furniture. The shapes of the bronze ritual vessels come mostly from pottery antecedents. Thus the bronzes are really complex translations of ceramics and other materials into metal, borrowing not only forms and designs from those other materials but, in the preparation of the molds for the bronzes, using some of the actual techniques for working them as well. Certainly motifs and forms in art styles are translated frequently from one material to another, as noted in the introduction. So they are too in Shang China. But in this case, the transformation is unusual in that it is accomplished by employing the *techniques* of the original material in making the translation into the new material; in short, the bronzes look as though they were carved ceramic or jade. This is not to say that the bronzes resemble contemporary works in those other materials, for carved ceramics and jades do not actually exist. Rather we are observing craftsmen so accustomed to particular modes of working, so acculturated in carving, incising, impressing, piece-meal technologies and apparently ignorant of lost-wax casting that their bronzes ended up looking as though they were carved and assembled. (We might note that in Sung times the reverse translation prevails: the exquisite celadons and porcelains are made to look like carved jade, cast bronze, and repoussé silver).[2] Clearly there are very strong cultural constraints at work here that impart this appearance to the bronze ritual vessels.

These bronze vessels show a strong balanced opposition of design which is, of course, inherent also in the opposition of positive/negative, model/mold and mold/metal. There is, moreover, a similar kind of *yin-yang* principle of opposition already apparent in the early oracle bones contemporary with the bronze castings (Keightley ms.). We also noted above that the ceremonial centers of the Shang sites were laid out in a basic two-fold opposed pattern; that there were two ruling houses alternating with each other; and that there were two worlds, one of gods and one of men, that had to be kept in balance with each other, a balance that was accomplished by ritual in which precisely these bronze vessels were employed.

These bronze ritual vessels are the products of a distinctive technological (and art) style characterized by balanced opposition, replication, translation, carving, impressing, and an emphasis on careful ceramic work, a style that is not found anywhere else in quite this manner and stands out as one of the hallmarks of early Chinese civilization. We have noted at the beginning of this section that the Shang foundrymen, along with other specialized craftsmen, appear to have enjoyed relatively favored positions in their societies. We assume that this is because they made objects of a prestige-reinforcing nature for the elite which was therefore willing to support these craftsmen at a high level. In return for this elevated socio-economic status, demands of high quality work were made on the craftsmen as these superb castings reveal. As has been noted elsewhere, the elites of highly stratified societies seem to require sumptuary goods to maintain their prestige and status, and will, as a result, support crafts of extraordinarily high labor intensivity with commensurately high technical virtuosity and quality (Lechtman and Steinberg In press). When the elite is not so small, nor so high-status differentiated, the crafts will tend to be of lesser quality, as we shall see in our third example below.

IRON AGE ANATOLIA

The kingdom of Phrygia in central Anatolia in the 8th century B.C. exhibits a class of bronze wine-serving and drinking vessels of exceptionally fine workmanship that are raised from hammer-forged sheets of metal intricately fastened together.[3] The place of origin of the unusual large cauldrons with siren handle attachments (Fig. 3) and their relations to similar pieces found elsewhere in the Near East, in Greek sanctuaries and in Etruscan burials has been much discussed (Herrmann 1966; Muscarella 1962, 1970; Young 1967). I do not wish to enter this argument here, for it is waged largely on the basis of iconographic analysis, poorly documented findspots, and strong cultural biases. I will concentrate instead on those vessels thought to be of local Phrygian manufacture and stress their workmanship rather than other aspects (Muscarella 1968:13; Young 1958:151). Two of the "imported" cauldrons with their accompanying "local" bowls and pitchers (Figs. 4–6) come from a well-recorded context at Gordion, in Phrygia, near modern Ankara (Young 1958:147ff.). It is because of good archaeological contexts that I shall focus on the Phrygian vessels rather than deal with the more spectacular, but far less well documented large cauldrons. It is just possible that this whole Phrygian industry is an example of diffusion from North Syria or even Assyria, but it is so close in time and place to the earliest extensive manifestations of this particular style of bronze working that we might well consider it indigenous to Gordion.

Gordion, the capital city of the kingdom of Phrygia which flourished especially in the 8th century B.C., lies in a large, well-watered plain at a ford of the Sangarios river. The plain appears to be sufficiently large and fertile to have supported a population of urban proportions. The main site of Gordion was occupied well back in the Bronze Age. Its important Phrygian levels of the 8th century B.C. are characterized by a monumental gateway leading to the civic center, which consists of an open plaza flanked by large, elaborately decorated public or official buildings (palaces, audience halls?). This large-scale building program which required centrally-organized planning, "urban renewal" and a concerted labor effort, is symptomatic of what we might expect from a powerful king who, Assyrian records inform us, was even able to resist the might of the Assyrian empire. We may assume that noblemen or other wealthy citizens were buried in some of the tumuli, that some kind of army protected the city walls, and that priests worked in some of the *megara* in the civic center; of bureaucrats we have less evidence so far. To judge from written sources, from the distribution of Phrygian goods elsewhere, and from the frequent occurrence of foreign goods in the local burials, Gordion had extensive trade connections with the Greek cities to the West and with the kingdoms of North Syria and Assyria to the South and

Figure 3. Large cauldron with siren ring-handle attachments from Midas Mound Tumulus, Gordion (now in the Archaeological Museum, Ankara). The Cauldron with its thick everted rim is raised from one sheet of bronze; it is about 2 mm. thick in section, about 75 cm. at its greatest diameter and about the same height.

East. Whether this trade was conducted by merchants or by the king and his diplomats is not clear.[4] So far no good evidence has been found of an agricultural peasantry which must have supported this kingdom. The fine crafts of potting, wood inlaying, ivory carving, timbered construction, and bronze working all appear, on the basis of stylistic analysis, to be indigenous, and they are surely the products of full-time specialists, probably supported by the king. Finally we should note that a part of Gordion's wealth may have been derived from the exploitation of local mineral resources evidenced both by the large number of bronzes in the tombs and by the unusually high zinc content of some of them which seems to indicate a special local ore. Midas' mythical golden touch is probably also a reflection of this mineral wealth.[5]

Figure 4. Round-bodied trefoil jug, from Midas Mound Tumulus, Gordion (now in the Archaeological Museum, Ankara). The body is made of one sheet, the trefoil mouth is made separately and attached with the ring at the neck; the strap handle is riveted to mouth and body. The jug is about 35 cm. in diameter and the same height.

Figure 5. Spouted one-handled bowl, from Midas Mound Tumulus, Gordion (now in Archaeological Museum, Ankara). Spout and handle are separately raised and attached to the thick one-piece body with small, carefully concealed pins. Bowl is about 20 cm. in diameter and 6 cm. high.

Figure 6. Petalled omphalos bowl, from Midas Mound Tumulus, Gordion (now in Archaeological Museum, Ankara). The omphalos and ridges around it appear to be cast and probably soldered into the bottom of this bowl which is otherwise raised and decorated by repoussée. The bowl is about 22 cm. in diameter and 7 cm. high.

About one kilometer north of the main urban center is an extensive cemetery of mound or tumulus burials which seem to span several centuries and contain funerary remains of extraordinary splendor. The burials would appear to belong to the royal family and wealthy noblemen, and are notable for the similarity of their contents. We will concentrate here on the largest and richest of these, but observe that very similar bronzework and pottery are found in several of the other burials. The largest one, called Midas Mound Tumulus by the excavator, consists of a chamber built of heavy timbers inside a giant mound of stone, earth, and clay (Young 1958: 147–51). Inside the chamber was buried a small elderly man laid out on a wooden bed, dressed in a garment fastened on the upper portion by "brass" fibulae. The tomb chamber was filled with resplendent gifts consisting of pottery, leather work, wooden tables, and

Figure 7. Cast fibula, from Midas Mound Tumulus, Gordion (now in Archaeological Museum, Ankara). The various balls and hemispheres were made separately and pinned (the smaller ones may have been soldered) onto the cast bow and clasp; pin and spring are hammered separately and then force-fit into the end of the bow. The fibulae from this tomb have an unusually low tin and high zinc content; this one is 5 cm. square.

screens, a bag filled with 145 fibulae (used perhaps as some kind of currency?) (Fig. 7) and 169 bronze vessels.

It is the vessels that interest us here. The excavator's list is as follows:

2 large cauldrons with siren ring-handle attachments (Fig. 3)
1 smaller cauldron with bull ring-handle attachments
10 round-bodied trefoil jugs (Fig. 4)
19 smaller trefoil jugs
2 spouted jugs
2 situlae
6 deep bowls with ring handles
4 deep bowls with bucket handles
16 shallow bowls with ring handles
3 rim-handled basins
2 spouted one-handled bowls (Fig. 5)
2 ladles
2 plain hemispherical bowls
7 ribbed omphalos bowls
37 plain omphalos bowls
54 petalled omphalos bowls (Fig. 6)

This would appear to be an elaborate banquet service mainly for the consumption of wine (most of the bowls are probably drinking vessels, especially the omphalos bowls with their raised central bosses which would afford a good grip to the banqueter). I assume their use as a wine service because of the excavator's comment about the poor state of preservation of the ceramics found inside the cauldrons which, he thought, may have been consumed over time by the wine in which they were left lying. I would suppose further that these bronze vessels played some role in the funeral ceremony, though they were also apparently used at some previous time to judge from the damage some of them sustained. We know from tomb finds in the Urartian region (e.g. Altintepe) and in Etruria (Vetulonia, Palestrina), that in those societies also, bronze cauldrons, along with serving and drinking vessels, form an important part of the tomb furniture of deceased people of considerable wealth (local chieftains?).[6] Cauldrons and some other bronze vessels are, incidentally, found dedicated at Greek sanctuaries such as Samos, Delphi, and Olympia where, to judge from Herodotus (IV, 152), they seem to be thank offerings of some sort. But whether this was their primary function or a later, secondary one, is not clear. In any case, the Gordion examples are bronze ritual vessels associated with a burial; they are made with consummate skill and care in a series of steps involving a number of different techniques. I am suggesting, moreover, that the smaller drinking and pouring vessels represent not only a distinctive art style that can be readily described and located in space and time, but that they also reveal a characteristic technological style all their own. The technological style, like the art style, is partly derivative from other Near Eastern areas, but essentially combines a number of features that we can call typically "Phrygian."

The smaller pitchers, drinking, and serving vessels are all notable for their very smooth, finished surfaces, their regular forms, and their painstakingly careful attachment of handles and spouts. In most cases the bodies of the narrow-necked or spouted vessels were raised from a single sheet of bronze, with spouts and necks attached either by rolling the body sheet over a ridge on the attached section especially made to accept it (Fig. 4) or by applying the attached portion with very small pins which were peened over and ground down so that no protruding heads remained (Fig. 5). Handles, on the other hand, were often attached with rivets which do show a characteristic form. The petalled omphalos bowls with high sharp ridges surrounding the central boss appear to have those unusual central portions somehow made separately and then attached to the center of the bowl, since the raised ridges on the interior do not correspond to indentations on the exterior (Fig. 6). The average thickness of the vessels is about 2 to 3 mm. and reveals absolutely masterful control of the forging properties of the metal. We see no traces of repairs of cracks or tears in the metal that can occur from inadequate annealing. The last stage of production was apparently a very fine abrasive polishing on both the inside and outside, which removed virtually every trace of hammer marks. The same skill of raising bronze sheet is apparent on the large siren-handled cauldrons with their thick everted rims which, for sheer size, regularity of contour and smooth finish, are even more impressive than the smaller vessels (Fig. 3). We suppose that though the cauldrons may well be imports and the smaller vessels local products, the underlying technological procedures and style are quite similar. The hallmarks of this style are meticulous raising of metal sheet, building vessels of composite forms by almost invisible means of joining sections, attaching parts (handles, spouts, etc.) with such fine pins that they are visible only upon the closest inspection, and abrading and polishing the vessels so that they show no trace of tool marks.

I have noted elsewhere (Steinberg 1970:103–6) and would like to observe again that in the early first millennium B.C., in Northwest Iran, North Syria, and Anatolia, there seems to have existed a metallurgical style of raised vessels with unusually high repoussé design manifest in situlae from Hasanlu and Gordion, cauldron stands and various drinking vessels found in Greek sanctuaries and Etruscan tombs, as well as in various poorly documented individual museum pieces all of which represent a distinctive artistic as well as technological style (see Muscarella 1970 for a complete listing). Essentially these pieces portray a vigorous pictorial style in very high relief, full of assorted Near Eastern motifs of hieratically opposed animals, combined animals ("Mischwesen"), trees of life, snakes, and in addition, there are whole vessels formed in the shape of animals' heads. The relief is generally so high that the interior of the vessel requires a smooth liner in order to be functional. There are only two of these repoussé vessels in the tombs of Gordion, both probably imported, but the high relief of the omphalos bowls is a notable local manifestation probably related to this general high repoussé style. Furthermore, in other deposits such as the Italic tomb-groups (Curtis 1919, 1925; see Muscarella 1970:117–21 for a list), the siren-handle cauldrons similar to those from Gordion are often found together with pieces of this repoussé style; in fact, stands made in this manner often support precisely these kinds of cauldron (Herrmann 1966). In at least three areas (Gordion, the Athenian Kerameikos, and Central Italy) these high repoussé vessels appear in funerary contexts; at Gordion and in Italy they also appear together with siren-handle cauldrons, and we might suppose they serve a similar ritual function in all these contexts.

The high repoussé vessels share with the Gordion raised, smooth-surfaced vessels a general tendency to extend by very elaborate hammering a single (or very few) sheets of metal. The epitome of this technique is undoubtedly the large cauldron with cast handle-attachments, which, it was noted, is found together with both the high repoussé and the smooth vessels. The cauldron, on the whole, shows greater technical affinity to the smooth, raised vessels of Gordion than to the repoussé examples from North Syria (or from wherever they may ultimately originate), though the cast handle-attachments on those same cauldrons show far-flung stylistic relationships (Muscarella 1962, 1968).[7] I do not wish to suggest any place of origin for those cauldrons but only to observe that they seem to share a technological style with the Gordion raised vessels.

I am suggesting that this mode of hammer-raising smooth bronze ritual vessels (the petalled omphalos bowls are exceptional) constitutes a distinctive technological style in central Anatolia during the early first millennium B.C., and it may well be related in some way to the high repoussé style from farther south and east which is both contemporary and earlier. Its relation to other crafts at Gordion is somewhat tenuous. Pottery shapes (especially such metallurgically awkward forms as the trefoil mouth and the long spout) appear to underlie some of the bronze vessel shapes, so that we have here, as in China, a translation of form from one material to the other. The very unusual, elaborately cast "brass" fibulae are not related technically (though the same high zinc content is found both in the fibulae and in some of the vessels), except that they also show a predilection for intricate workmanship without revealing any traces of the processes involved (Fig. 7). The joining of the various sections and adornments of these fibulae is as well hidden as are the means of spout and handle attachment on so many of the Gordion vessels. The intricate wood carving and inlaying revealed on the furniture in the Midas Mound Tumulus is also a notable craft at Gordion, but shows little relation to the metal working. That potters and bronze workers saw each other's work has been noted above, but it is often difficult to determine which way the influence moved. We cannot, in short, speak of a general Phrygian technological style, but of a style of bronze working distinguished by the features noted above which, to some extent, is imitative of pottery prototypes. What all the crafts do seem to share though is an extraordinary degree of control by the workman over his materials resulting in high quality work supported, we must assume, by a small elite for the reinforcement of its prestige and power. Again, as in Shang China, this patronage had certain aesthetic and technical canons, but in Phrygia they appear to be expressed somewhat differently in each craft and material. This lack of a single style, as coherent and integrated as the Chinese, may be due either to influences from neighboring cultures (in fact, I did suggest above that there were close trade relations with some of those neighbors), or perhaps to an internal lack of cultural integration for which we do not, as yet, have any other evidence.

BRONZE AGE CENTRAL EUROPE

The third technological style that I wish to examine is a bronze sheet hammering and riveting technology that appears to be native to south-central Europe, on the Danubian plain, in the 13th and later centuries B.C., a product of the so-called Urnfield and Hallstatt cultures.[8] The find-spots of these vessels are not sumptuous inhumation burials, as in the previously discussed cultures, but either cremation burials (in which the material is all gathered together and placed in a pit in the ground) or hoards of scrap bronze destined for reuse. There are a few richer princely tumuli burials containing similar vessels, but these too are in contexts of cremations buried in pits under the tumuli. Prehistorians seem to agree that the beginning of the Urnfield period, in which these bronzes originate, is an unsettled time, marked by large-scale migrations from north-west to southeast central Europe. These migrations seem to have brought this new technology into the Hungarian Plain near the metal-liferous Carpathian Mountains where it resided for a few hundred years before spreading West and North again, and then South into Italy under the energetic Hallstatt traders (Gimbutas 1965:328ff; Merhart 1952:58–61).

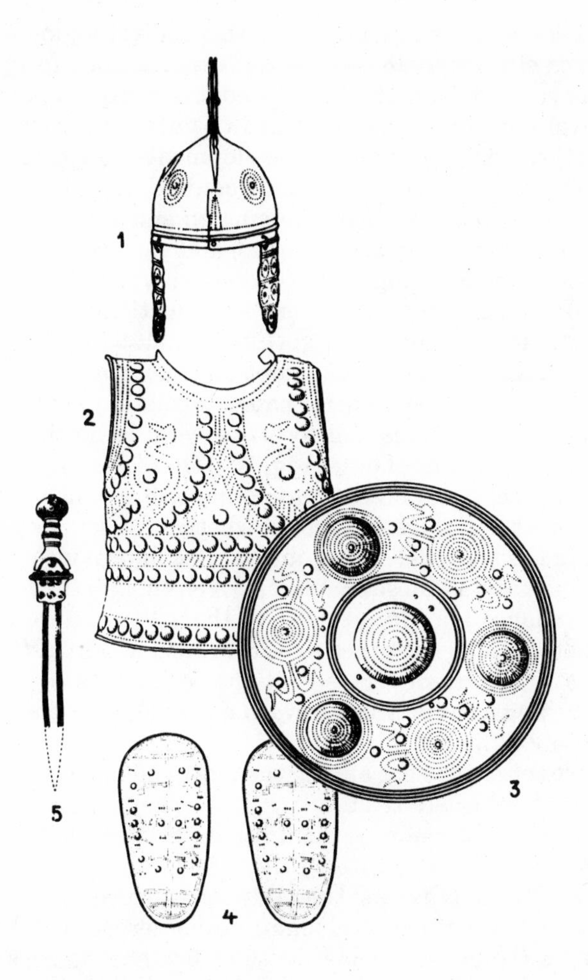

Figure 8. Body armor from various European sites put together to show the nature of a warrior's defensive ensemble. (From Merhart 1952:pl. 26).

The Urnfield culture in which this technology thrived is characterized as a thinly stratified society consisting of an agricultural-pastoral peasantry, craftsmen, and a powerful warrior aristocracy led by a chief or prince of sorts. There may have been a small class of traders and possibly a separate priesthood.[9] Settlements of about 200 souls were small and heavily defended; large towns were clearly not a characteristic settlement pattern of this agrarian, warrior culture which preferred a more dispersed rural form, though concentrating its dead in large cemeteries or urnfields full of pots containing cremation burials and grave goods (Piggott 1965:140–1; Gimbutas 1965:301 ff.).

The rich copper and tin resources in this central region of Europe had been exploited since early in the Bronze Age. The chief form of metal production was casting, and the main products were weapons. The major finds of metal objects generally occur in or near the metal-working centers around these rich ore sources. Development of an effective new slashing sword during the Urnfield period required also the development of new armor which was made from hammered and annealed bronze sheets formed into helmets, cuirasses, greaves, and shields (Fig. 8). A by-product of this sheet-hammering armor industry, which thrived particularly in the Danubian plain, was an elaborate production of thin sheet buckets (Fig. 9), cups, and strainers. Piggott argues (1965:154) that these indicate the introduction of wine consumption to central Europe. There are a number of variants of these basic shapes consisting of one- and two-handled cups of different sizes and forms, several larger bowls or pitchers with pail handles, and amphorae (Fig. 10). These vessels might well be intended for wine-service, and they are found either in hoards or in graves. Their function as burial-ritual vessels, whether for drinking at the funeral banquet or in the afterlife, seems quite sure and provides a good analogy to our Phrygian and Chinese examples. Even though the function may be similar, the shapes and decorations of these vessels are different from those characteristic of the other two traditions we have studied. What is even more striking, however, is that the techniques by which they are made and the apparent attitude toward materials that they represent are also completely different.

The buckets or situlae are among the most elaborate and interesting of these vessels (Merhart 1952:29–38) (Fig. 9). They range in size from 17 to 50 cm. high and are made by hammering bronze sheets to a uniform thickness of about 1 mm., cutting them to shape, then joining them by riveting 2 sheets together for the body and a third one for the bottom. The bottom sheet is turned up along its circumference to a height of 2 to 3 cm., forming a flange that holds the sides firmly. The necks and shoulders are strengthened by a number of corrugations, and the rims are reinforced by rolling them over a hoop of bronze or iron. The base is often reinforced by various straps and plates. Thick strap handles are riveted on the shoulder and inside the rim. Repairs are also made by riveting thin sheets over holes or tears. Apparently hammers of horn or wood rather than of stone or metal were used in their production, since hammers of softer materials seem to have been responsible for the pleating of the metal on the base sheets of these buckets (Tylecote 1962:146).

Using uniformly-hammered, thin sheets cut to shape and fastened together by rows of densely spaced rivets is certainly a different and easier way of making large containers for liquids than either casting them or raising them from one piece. The flat-bottomed shape (like modern pails)

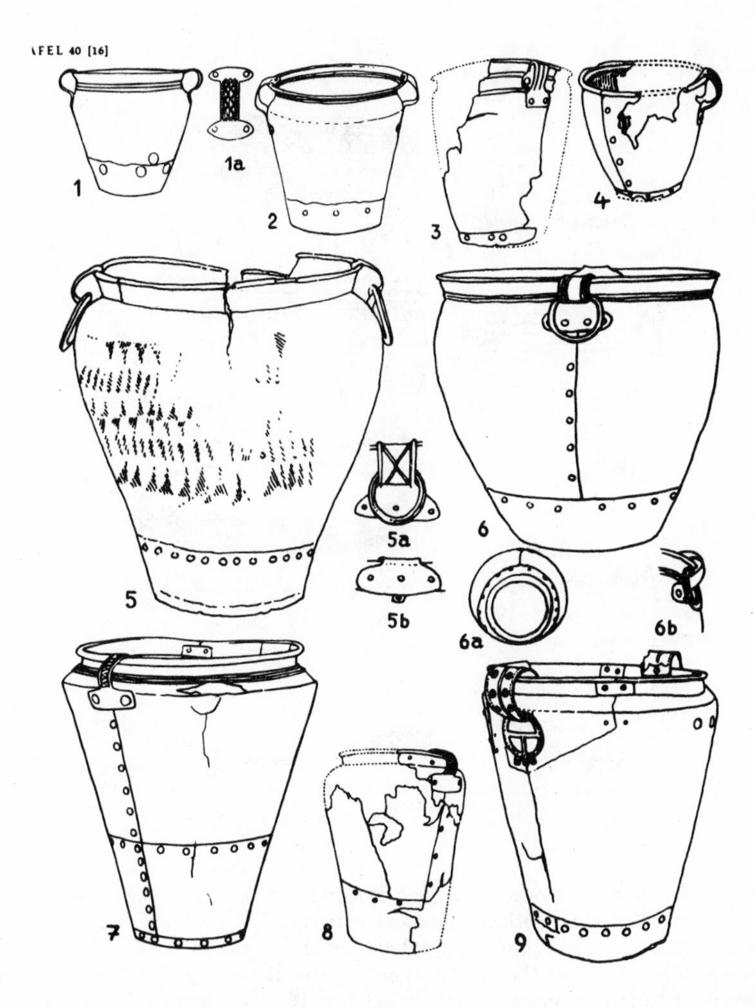

Figure 9. Various buckets of the Kurd type made of hammered sheets riveted together with strap handles attached separately, also by riveting. No. 6 is about 40 cm. high, the others are to the same scale. (From Merhart, 1952:pl. 16).

eliminates the need for separate stands or for feet, both of which were probably important considerations in their design.[10] But the flat bottom (often requiring reinforcement with extra plates and straps) which meets the sides at almost ninety degrees requires a special mode of production—it would be very difficult to raise walls at such a steep angle from the base. Hawkes and Smith (1957:181) insinuate that this relatively simple sheet-and-rivets technique was an outgrowth of the technical inferiority of European smiths, and the ease of production by this technique may well support this view. It should be added that the work is not

Figure 10. Pail-handled bowls, on the left, and cups, on the right. The bodies are made of single sheets of thin bronze with the handles riveted on. The scales are different: bowl no. 1 on the upper left is about 28 cm. largest diameter and 15 cm. high, the other bowls are to the same scale; cup no. 4 on the right is 24 cm. largest diameter and about 14 cm. high and the other cups are to the same scale. (From Merhart 1952:pl. 2 and 8).

as carefully controlled and finished as in the other two styles we have discussed. A particularly interesting phenomenon is that this techinque, which begins on the buckets of Kurd and Hajdu Böszörmény type, is then transferred to other shapes because of its relative practicality and ease of execution.[11] When Mediterranean-type cauldrons (like those discussed for Phrygia) are imitated as far afield as Britain they are made in the local European technological style of sheets and rivets rather than raised from one piece (Hawkes and Smith 1957; Tylecote 1962:pl. XII). This is a striking example of the persistence of a technique even when a new form is introduced.

The various smaller bowls, cups, and pitchers (for serving and drinking the liquid carried in the buckets?), which usually have rounded bottoms, are raised from one sheet with the handles attached, again, by extensive riveting (Merhart, 1952:1–28) (Fig. 10). Some of the smaller vessels show occasional attempts at making offset or flat bottoms, but in these cases the walls rise in a gradual arc above them, not abruptly, as on the larger pails; thus such vessels did not require piecing together. It is interesting to note again that where the shapes are complicated, as on the so-called amphorae, the vessels are built up out of preformed sheets that are riveted together. In fact, the rivets came to take on an important decorative function in addition to a purely mechanical one, and developed stylized conical forms in later examples (this is also true in the Anglo-Irish tradition).

Finally, before leaving this distinctive Central European vessel-building technology, it should be noted that the secondary decoration added to these vessels is also unusual and culture-specific. It falls essentially into

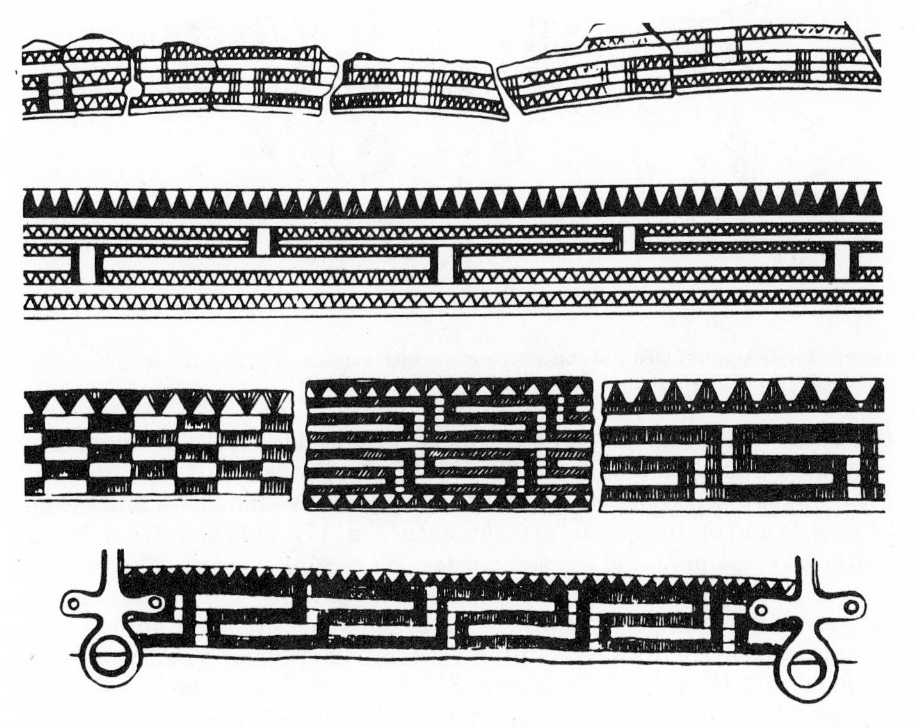

Figure 11. Hatched geometric designs made probably by engraving rather than tracing, near the rims of various vessels. (From Merhart 1952: pl. 7).

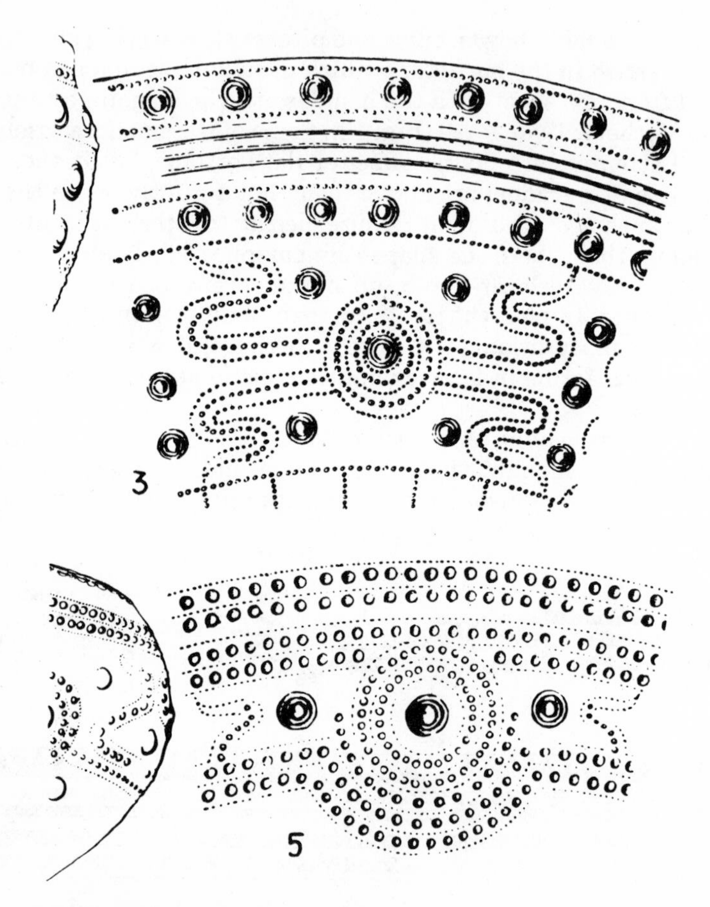

Figure 12. Designs of bird and sun motifs made with various sizes of punches from inside the vessel. (From Merhart 1952:pl. 23).

two types which I do not think can as yet be chronologically or regionally separated (Merhart 1952:7–12, 38–58). The first consists of very fine, engraved, angular geometric patterns used along the shoulders and necks of vessels and on the handle attachments (Fig. 11). This appears to be an imitation of painted and incised borders on ceramic vessels. The second type of decoration is equally distinctive and is found more often on the bodies of the vessels. It consists of designs representing animal and rounded geometric forms produced by dense clusters of punch marks of various sizes (Fig. 12). This punched decoration tends to be found more frequently on the buckets than on the other vessel shapes. It is from these same buckets that the "situla art" of the 7th and succeeding centuries developed in which Hallstatt shapes and decorative techniques were combined with newly introduced motifs and repoussé working, both of which came from Greece and the Near East (Arte delle situle 1961).

We have here then an indigenous technique of making and decorating bronze ritual vessels with its own internal development distinctly different from the other traditions we have examined. It may have been prompted, if Piggott is right (1965: 169), by foreign influences in terms of viticulture and wine-consumption, but it certainly developed into a bronze-working technology all its own. What is particularly striking about it is that when vessel forms or decorative schemes (both in terms of motifs and arrangement on the vessel) were introduced to these cultures from elsewhere, the bronze smiths persisted in doing things in the manner to which they were accustomed, not in the new, foreign way. Once again, we can speak of a persistent technological style.

The relations of this bronze sheet and metal work to other features of Urnfield culture are still tenuous. Ceramic wares are rather coarse, burnished, monochrome, and incised and appear to be fired at low temperatures, often under reducing conditions and probably in open fires. Other crafts are all rather poorly preserved, though we suspect that there was extensive leather and textile working, and we do know of some good, if rather rustic, gold smithing. Wagons and chariots of wood with bronze fittings and horsetrappings of bronze were also made. War, apparently, was a way of life, so that good arms and armor were essential. Exploitation of the local copper and tin resources was necessary for the manufacture of the bronze ornaments that embellished such paraphernalia. It seems that metalworking was concentrated in these mining regions, as evidenced by hoards of both cast and beaten bronzes. Cast offensive weapons and tools appear in huge numbers and are, along with defensive armor of sheet metal and the related fine festive vessels, the main products of metallurgy (Gimbutas 1965:307–28). Gimbutas asserts that most of the metalworking was done at centers near the mines, which suggests that it was carried out by a class of specialized, skilled craftsmen. This may be true of the casting of weapons, but the sheet-metal work which produced vessels by riveting together prehammered and precut sheets lends itself to an itinerant tinker form of production. The wide variation in shape and detail also suggests such an individualized form of craft production. In fact, there is generally a kind of imprecise, rustic, even crude, quality to the work which may well be commensurate with the relatively lowly stratified chiefdom stage of socio-cultural integration found in this culture.

CONCLUSION

I asserted at the outset that different cultures have distinctive local methods of working materials which may be called technological styles. I think that these styles represent predilections for working materials that are at once purely technical—relating to the mechanical properties of materials—but are also related to aesthetic and other cultural constraints such as function, socio-economic structure, and the like. In the three examples discussed I have tried as much as possible to keep material and vessel function constant and, where possible, have used cultures with similar forms of socio-cultural integration, though I am not sure that I have been successful in that latter regard. Certainly the craftsmen producing these objects, whether in proto-imperial Shang China or in the kingdom of Phrygia, were full-time, highly skilled specialists patronized by powerful ruling elites, be they imperial or royal families. The Central European chiefdom's craftsmen, on the other hand, may represent an aberrant example in that they were either travelling tinkers making vessels to order for local warriors, or possibly specialists localized near ore sources and away from the settlements where their ultimate markets lay (Gimbutas 1965:307). But the findspots of all these vessels, as best we can determine, were precisely in the burials of the patrons of these crafts, which is to say that the objects were made by craftsmen who probably worked for the people who used them (again, the European example is ambiguous). The objects, we think, were deposited in the contexts for which they were originally produced, namely burials. If this is so, it gives us but one more element in the close fit between technology, craftsman, product, patron, and deposition. Another element of similarity, or rather of comparability, is that each of these technologies has apparently developed quite independently of outside influences. We are fairly certain in the case of China and Europe, and probably even of Phrygia (granting its close relations to North Syria), that we are viewing these particular forms of technologies near the beginning of their development, before they have become too diffused through outside technical or artistic influences. Furthermore, although in all cases the basic techniques were probably imported, I believe that the manifestations I have discussed developed on home ground and were not brought from elsewhere.

With so many of the external relationships reasonably constant, what conclusions can we draw about technological styles? Each culture has developed its own particular way of fashioning ritual bronze vessels (ritual is used here to mean some kind of drinking ceremony associated with funerals or life-after-death burials). Furthermore I suspect that the reasons for the specific technical forms employed are to be found in other aspects of each culture, especially in the standards of appreciation, the general level of skill and even the techniques used in other technologies, but also in such features as the culture's adaptation to its environment, its subsistence patterns, religion, and trade. Of necessity I have refrained from detailed analysis here of the latter features, but I can say something about the attitudes of these various cultures to materials and about their means of solving technical problems.

The Chinese bronze workers were, first of all, foundrymen, not smiths;[12] they consciously imitated earlier ceramic shapes in bronze and did so by lavishing particular attention on the ceramic molds in which their bronze imitations were cast. The carving, incising, impressing, burnishing, and adding bits of design are all techniques shared with the potters, and were all performed in the "ceramic", first stage of the production of these bronzes. Solutions of undercutting, projection, and joining extraneous parts were all met by intricate methods of mold building, not by original design on the model, nor in wax, nor by subsequent work on the finished cast bronze. When all is said and done the Chinese vessels look as though they might just as well have been *carved* out of clay, wood, or stone rather than made in bronze—but they weren't! The new material, bronze, was so beautiful, so strong, and so intriguing—to melt a solid metal and constrain it to take a new shape is indeed intriguing—that it had somehow to be used for this important ritual purpose by the Shang dynasts. (It is furthermore interesting to observe that the *yin-yang* balanced opposition that comes to play such an important role in Chinese metaphysics is reflected in that positive/negative, shaper /shaped mode in which the foundryman constantly must have been working and thinking. Is it too extreme to suggest that the development of that concept was related to this technology?) The vessels are veritable technical *tours-de-force* in that they skillfully adapt metal to an essentially carving technique. This is a technological style that is distinctively Chinese, that persists for about 1000 years and is then replaced by Western lost-wax casting which completely changes the nature and quality of the bronze industry. There is, incidentally, a decorative style of various recurring, balanced, and opposed motifs that is associated with this technology which also gradually changes as the mold-building technique changes. Finally I should observe that the complex sectional-mold technique was basically limited to ritual vessels; weapons were cast in simpler bi-valve molds, and later rounded vessels, decorated with inlay, were executed in repoussé; incense-burners and belt-hooks were all later introductions produced by the lost-wax method. This is what I mean by the limited repertoire of particular technologies. Certainly other objects of every day use might have been cast by this same method, but they were not; sectional-mold casting of bronze was largely limited to the production of ritual vessels.

In the Near East I focused on a technique of finely finished, smooth raising that culminated in the production of large bronze cauldrons. There is no question that the technique of raising vessels from sheet hammered out of thick billets of metal goes back at least to the early third millennium in the Near East, but making them so smooth, so symmetrical and with such well-hidden joins is a characteristic of the time and place discussed here. Furthermore, the attachment of handles and spouts by practically invisible means seems to have been an important part of the conceptualization of this technology. Goals here appear to be smoothness, quality of finish, hiding of any traces of working—in short, an aesthetic of pure, clean form that is not often encountered in ancient bronze working. There is also a rejection of seams, attachments (except for the cast and exquisitely tooled siren handle attachments of the cauldrons), or even decorative elements that detract in any way from the appreciation of the form itself. Where decorative elements are adjoined, such as the floral petals on the omphalos bowls, they are executed with a cold regularity and precision that reduces them to pure bravura repoussé elements rather than real designs. This is vessel raising at its finest: the craftsman forms vessels by making sheets of extraordinary uniformity and evenness in relatively simple forms which, interestingly enough, must either be held in the hand or supported on stands, for their bottoms are round, not flat. The vessel form of evenly hammered bronze lies at the heart of this technology.

The Phrygian metal-raising tradition was associated with other refined crafts of this small kingdom of central Anatolia, but I am not yet able to show how this particular tradition relates to Phrygian culture, though I hasten to add that little enough is known about the culture in general. Careful, meticulous, labor-intensive if rather simple work is typical of Phrygian masonry, mosaic, planning, tumulus building, carving, potting, and so on. This is not a great cosmopolitan trading city with far-flung diplomatic and commercial relations, though at Gordion we do note some foreign influences. On the whole, it was rather a small principality peripheral to the Assyrian power to the east, to the re-emerging Greek cities of western Asia Minor, or even to the overpowering steppe nomads like the Kimmerians and Scythians who were eventually to overrun and conquer Phrygia. But it was an important buffer kingdom that seemed to play a leading role in the complicated balance of power of 8th century Western Asia. One can imagine then that the Phrygian patron of craftsmen was desirous of "keeping up with his neighbors," of presenting the image of a petty dynast of wealth, power, and prestige, who needed quality goods to celebrate his life after death. But this only argues about good craftsmanship, it does not tell us why metal was raised rather than cast, why the particular shapes and means of attaching handles and spouts were used, why there developed a predilection for the beautiful abstract forms—we need to learn far more about other aspects of Phrygian culture before we can go further in this analysis.

Finally, a concluding word on the Central European style of bucket and vessel forming. There is a certain rustic practicality in the flat-bottomed buckets and in the flat bases and ring feet of some of the other vessels—they were meant for use and they were to be set down on floors or on tables. The same rustic practicality is revealed in the rather poor quality of the work: edges are not always trimmed, rivets are not always symmetrical and regularly aligned, and certain short cuts are used such as rolling the rim of the buckets over a preformed hoop rather than thickening the metal of the rim as on the Phrygian cauldrons. Vessels are sometimes a bit lop-sided; pleats appear in the thin metal where sheets were gathered in too hastily; tears occurred where too little annealing was employed. There is something crude about this technology that bespeaks the tinker rather than the high-level court artisan. The Central European vessels were made in large numbers, possibly even mass-produced, and were not necessarily made for a supporting patron, unlike the situation suggested for Phrygia or China. Arms cast in bronze certainly were being mass-produced for this "warrior aristocracy," and it is just possible that similar mass-production was required for the grave goods. It is not that the tinkers were incapable of better work, as an occasional bucket or situla indicates, but they served a larger and probably less selective or quality-conscious clientele than their eastern counterparts. Furthermore, they were also called on to make armor, in addition to bronze banquet-ware, which required them to find more expedient solutions to technological problems; they opted for prehammered and cut sheets and rivets. The socio-economic situation of the tinker in the Central European chiefdoms, and the demands made on him by his culture, may have been quite different from those placed on his eastern counterparts who were probably working for more aristocratic elites. I suggest that we see some of this difference in the reduced quality-control of these practical but hasty products when they are compared with the beautifully formed, carefully finished, and somewhat less practical works of the Phrygian and Chinese craftsmen. These are different cultures, with slightly different social structures, and with different criteria of craft excellence, producing three distinctive technological styles.

The difficulty of finding closer parallels among these three technologies or between metal working and other cultural features raises an interesting question. At the beginning of this paper I discussed how completely a technology is integrated into the cultural matrix in which it thrives. Yet there are instances when that integration is not so clear. Diffusion of techniques from elsewhere, desire of the patron (or craftsman) to imitate foreign crafts, and general lack of cultural integration due to various processes of change are all possible reasons for this incongruence. It is also possible that the strength of the integration between a feature such as technology and other subsystems of a culture is a function of evolutionary complexity; that is, that internal parallels among subsystems may be less evident in more complex, variously stratified and segmented societies. This is an hypothesis that requires very careful study, however.

I have tried, in the foregoing study, to show how three cultures in distinctly different environments, with different cultural traditions and institutions but with certain similarities, such as an agrarian subsistence base and specialized metal workers, produced bronze ritual funerary vessels in three completely different techniques. I have also tried to relate these technological predilections to other features in the cultures such as their other technologies, aesthetics, or, in the case of China, a metaphysic of opposition. But we can only guess at the aesthetic sensibilities, religious beliefs, economic patterns, and intricacies of social organization of Phrygia and the Urnfield cultures. As is so often the case with archaeologically known complex societies, much of this important information is missing and can, at best, only be inferred. For this reason it is helpful to observe technological styles in literate societies or in cultures that can be studied ethnographically. But what I hope I have succeeded in doing is to suggest that technological styles exist, that technologies have distinctive ways of interacting with the cultures in which they are imbedded, and that this is a worthwhile field to study in order to learn more about man's varied ways of adapting to his environment.

NOTES

[1]This paper has gained immeasurably in accuracy and sharpness by the patient reading and comments of Louise Cort, James Howe, Jean Jackson, Heather Lechtman, Oscar White Muscarella, Jane Scott, Cyril Stanley Smith, and Ruth Tringham. I am very grateful to all of them for the time that they have taken to aid me in making this paper more comprehensive and useful.

[2]The beautiful green Northern Sung celadons decorated with incised and molded floral patterns look very much like jade. The fine white porcelains and porcelaneous stoneware, like Ting and Ching pai wares with their subtly carved floral decorations, look very much like imitations of repoussé silver bowls.

[3]I would like to dedicate this section about Phrygia and the metalwork of the Midas Mound to its excavator and my mentor—Rodney S. Young—who aroused my interest in bronze work of this period. Professor Young was killed in an automobile accident in the fall of 1974 before he could publish most of this magnificent material. I hope he would not have been too upset by my treatment of his Phrygians and their technologies.

[4]Whether an independent merchant class or the king's own diplomats are conducting long-distance trade is a problem of considerable importance to our understanding of the social structure of ancient societies; this whole issue is treated well by K. Polanyi (1957:12ff).

[5]Several scholars, including Rodney Young the excavator, have suggested ingeniously that the gold-colored brass produced by the high zinc content of the copper alloys may be responsible for the "golden touch" story.

[6]There is an extensive literature on tumulus-mound burials in Europe and Central Asia well summarized by Piggott (1965) and Gimbutas (1965).

[7]The place of origin of these cauldrons is much discussed by archaeologists and art historians interested in this period of Near Eastern and Greek art. It is much to his credit that Oscar White Muscarella was probably the first to argue cogently that this origin is to be sought in North Syria.

[8]We are here on much looser chronological footing than in the other two culture areas since the chronologies in Central Europe have been constructed on the basis of stylistic (frequently subjective) typologies of artifacts and on correlations with finds of better-datable Aegean material which is all a bit shaky; Renfrew's (1970) new examination of these chronologies bears somewhat on this problem.

[9]Gimbutas (1965:310) describes a more highly stratified society consisting of a king or chief, a village headman, eminent people, a laboring class of well-to-do and poor, and prisoners, but then she asserts it is not a class society! What appears to be lacking is a powerful elite.

[10]The Greek Geometric tripod cauldrons offer one interesting solution to this functional problem: legs are riveted (or occasionally cast-on) directly to the round-bottomed vessel.

[11]A similar sheet-and-rivet technique is found on "copper" vessels in the Shaft Graves of Mycenae and on occasional Minoan relatives (Catling 1964: 166–89). These apparently precede the European examples in time, but with the new questions raised about the European chronologies we cannot be sure which way the influence traveled.

[12]"Smith" generally refers to a craftsman who forges metal, such as a blacksmith, whereas "foundryman" is used to describe a founder or caster of metal.

LITERATURE CITED

Arte delle situle
1961 Mostra dell' arte delle situle dal Po al Danubio. Padova: Florence, Sansoni.

Barnard, N.
1961 Bronze Casting and Bronze Alloys in Ancient China. Canberra: Australian National University.

Catling, H. W.
1964 Cypriot Bronzework in the Mycenaean World. Oxford: Oxford University Press.

Chang, K. C.
1974 Urbanism and the King in Ancient China. World Archaeology 6, 1:1–14.
1968 The Archaeology of Ancient China. Revised edition, New Haven, Conn.: Yale University Press.

Chêng, Tê-K'un
1974 Metallurgy in Shang China. T'oung Pao LX, 4–5:209–29.

Curtis, C. Densmore
1925 The Barberini Tomb. Memoirs of the American Academy in Rome 5.
1919 The Bernardini Tomb. Memoirs of the American Academy in Rome 3.

Gettens, Rutherford J.
1969 The Freer Chinese Bronzes, II—Technical Studies. Washington, D.C.: Freer Gallery of Art.

Gimbutas, Maria
1965 Bronze Age Cultures in Central and Eastern Europe. The Hague: Mouton.

Hawkes, C. F. C. and M. A. Smith
1957 On Some Buckets and Cauldrons of the Bronze and Early Iron Ages. Antiquaries Journal 37:131ff.

Herrmann, H. V.
1966 Die Kessel der Orientalisierender Zeit, I (Kessel-attaschen und Reliefuntersätze). Olympishe Forschungen VI, Berlin: de Gruyter.

Keightley, D. N.
Ms. Shang Metaphysics. Paper given at the Association for Asian Studies Meeting in March, 1973, Chicago.

Lechtman, Heather
1973 The Gilding of Metals in Precolumbian Peru. *In* Application of Science in Examination of Works of Art. William J. Young, ed. Pp. 38–52. Boston: Musuem of Fine Arts.

Lechtman, Heather and Arthur Steinberg
In press The History of Technology: An Anthropological Point of View. *In* Proceedings of the International Symposium on the History and Philosophy of Technology, University of Illinois at Chicago Circle, 1973.

Loehr, Max
1968 Ritual Vessels of Bronze Age China. New York: Asia House Gallery.

Merhart, G. von
1952 Studien über einige Gattungen von Bronzegefässen. Festschrift des Römisch-Germanischem Zentralmuseums. Mainz, II:1–71.

Muscarella, O. W.
1970 Near Eastern Bronzes in the West: The Question of Origin. *In* Art and Technology. S. Doeringer, D. G. Mitten, A. Steinberg, eds. Pp. 109–128. Cambridge, Mass.: M.I.T. Press.
1968 Winged Bull Cauldron Attachments from Iran. Metropolitan Museum Journal 1:7–18
1962 The Oriental Origin of Siren Cauldron Attachments. Hesperia 31:317ff.

Piggott, S.
1965 Ancient Europe. Chicago: Aldine.

Polanyi, K., C. Arensberg, H. Pearson
1957 Trade and Markets in the Early Empires. Glencoe, Ill.: Free Press.

Renfrew, Colin
1970 New Configurations in Old World Archaeology. World Archaeology 2:199–211.

Service, Elman
1975 Origins of the State and Civilization. New York: W. W. Norton.

Smith, Cyril Stanley
1972 Metallurgical Footnotes to the History of Art. Penrose Memorial Lecture, Proceedings of the American Philosophical Society 116 No. 2:97–135.

Spier, Robert F. G.
1970 From the Hand of Man. Boston: Houghton-Mifflin.

Steinberg, Arthur
1970 Introduction to Part 2. *In* Art and Technology. S. Doeringer, D. G. Mitten, A. Steinberg, eds. Pp. 103–106. Cambridge, Mass.: M.I.T. Press.

Treistman, Judith M.
1972 The Prehistory of China. Garden City, N.Y.: Doubleday.

Tylecote, R. F.
1962 Metallurgy in Archaeology. London: Edward Arnold.

Watson, William
1962 Ancient Chinese Bronzes. Rutland, Vt.: Tuttle.

Wheatley, Paul
1971 The Pivot of the Four Quarters: A Preliminary Enquiry into the Origins and Character of the Ancient Chinese City. Edinburgh and Chicago: Aldine.

Young, Rodney S.
1967 A Bronze Bowl in Philadelphia. Journal of Near Eastern Studies 26:145ff.
1958 Gordion Campaign of 1957: Preliminary Report. American Journal of Archaeology 62:139ff.

4

The Role of Primitive Technology in 19th Century American Utopias*

MARK P. LEONE

University of Maryland, College Park, Maryland

This essay is concerned with four problems in analyzing 19th century American utopias. Why did they choose to foster primitive technologies? How is this choice related to the social problems they sought to solve? How did primitive technology guarantee them short lives? And, what does the style of their technology, as opposed to its function, tell us of the utopias' fundamental relationship to the United States? These may seem like four rather disparate questions, but they are linked in this attempt to correct two prominent misconceptions—rather, skewed emphases—regarding our utopias in the last century. The first misconception portrays

*Professors Boon, Elliott, Schuyler and Sorenson all read the first draft of this paper and made very careful comments on it which resulted in quite substantial improvements. I am very grateful to them for their willing help.

utopias as attempting independence and separation and ignores utopian desires to be a vanguard to America. This view thus avoids highlighting the reciprocal ties between the United States and its utopias. The second misconception regards the short existence of most 19th century utopias as an inexplicable misfortune, rather than as a structural correlate of both the technological base they deliberately sought and of deriving their identities from the parent society's problems. Some light can be shed on these misconceptions by trying to answer the first four questions.

I would like to begin characterizing utopias by discussing their isolation, a condition they all sought. The famous 19th century examples: Oneida, Amana, the Shakers, and Mormons; the earlier Pilgrims and Puritans in their religious commonwealths; and today's communes all seek physical separation from the society they are reacting against. Removal is so characteristic of utopias—it is common to both the successful and the unsuccessful (Kanter 1972:92)—and, in fact, it is part of a sociological definition of what a utopia is. Consequently it seems appropriate to ask: what have they isolated themselves from? And what is the upshot of the isolation?

Most utopias chose isolated locations as a way of managing their environments, including all aspects of the world impinging on them. In addition to closing out the world they sought institutional completeness as a way of guaranteeing their self-sufficiency. They usually sought an agrarian existence in a thoroughly planned community and consciously thought out plans for most of the utilitarian spheres of life, including many of the items of material culture they used to maintain self-sufficiency. This isolation plus their self-sufficiency led to the creation of a built environment in which many items had religious significance.

Fences, lines of trees, houses, town plans, furniture, clothing, and so on all had meanings and sentiments attached to them showing that they clearly functioned as much in the domain of religion as in economics and subsistence. The actual process of manufacturing furniture was one of the principal acts of Shaker worship. Building dams, planting trees, and plowing all involved acts of Mormon worship which endowed the implements used and products reaped with religious meaning. Neither Shaker furniture nor Mormon irrigation canals were holy objects to be venerated; they were not direct manifestations of the supernatural. Rather they were objects which operated in the everyday world but which also, like a set of prayer beads, simultaneously facilitated worship. In this sense much Shaker and Mormon technology—really, for present purposes, material culture—was sacred technology, devices that allowed access to the sacred.

Thus sacred technology and isolation are two linked characteristics common to American utopias. Sacred technology comes out of striving for institutional completeness which is in turn a part of isolation, and all stems from deliberate, concerted efforts to invent a separate and unique environment. These two characteristics are important in this paper because they allow an exploration of the novel relationship between economics and religion which exists when a group uses religion to isolate itself and to limit its access to technology. Since this paper is about the technological base for utopias, it must consider isolation and sacred technology, but it examines them primarily to see the curious relationship between our utopias and American society as a whole. Utopias could not operate without these two traits, but with them they were and are harnessed to America.

To begin to build an explanation for the related problems of utopian isolation and sacred technology I will use some of Roy Rappaport's analysis of the connection between ritual and ecological regulation. My aim in this is, of course, two-fold. One is to clarify the role of utopias in American society; the other is to demonstrate and to extend the utility of Rappaport's model especially as it concerns primitive technology.

Rappaport suggests that in technologically simple societies the sacred will regulate key subsistence variables. This means that in the relative absence of power, religious power, that is, power derived from imaginary (unempirical) sources, will be used to maintain ecological and economic balance.

> I have argued . . . that in technologically simple societies . . . the sacred and numinous form part of an encompassing cybernetic loop which maintains homeostasis among variables critical to the groups' survival. (Rappaport 1971:39).

Further, the sacred and numinous operate through religious ritual.

> The virtue of regulation through religious ritual is that the activities of large numbers of people may be governed in accordance with sanctified conventions in the absence of powerful authorities or even of discreet human authorities of any sort. (Rappaport 1971:38). Whereas most men are willing to accept such axioms as 'the shortest distance between two points is a straight line' as the basis for some of their behavior, they are likely to be more dubious about accepting calls to fight in distant wars or production quotas whose rationale they do not understand or believe to be in their own interest. (Rappaport 1971:36).

In the absence of sufficient coercive power to enforce important decisions, sanctity can be used to accomplish the task. This can occur in societies which have simple technologies because of their evolutionary status, or in societies which have simple technologies because, like American utopias, they have deliberately chosen to manipulate their environments to exclude complex technology or to place themselves beyond its reach.

In such technologically simple societies authority to decide and to regulate stems from the use of "ultimate sacred propositions. . .", and the effectiveness of the latter depends on how well regulated material conditions are.

> Inasmuch as the religious experience is an intrinsic part of the more inclusive emotional dynamics of the organism, which are closely related to its physical state, it is at least plausible to assume that religious experiences are affected by material conditions. But in technologically undeveloped societies the latter are at least partially a function of the control hierarchy that the religious experience itself supports. It may be suggested that the willingness, indeed the ability, of the members of the congregation to affirm through religious experience the propositions that sanctify the control hierarchy may be in some degree a function of the hierarchy in maintaining homeostasis in and among those variables crucial to the congregation's survival. (Rappaport 1971:39–40).

This says that the more successful a religion is at regulating empirical conditions, the more believable its unempirical statements will seem, and the more believable these are, the more effective their capacity to control reality will be.

Rappaport also says that sanctity is degraded by power.

> When, because of technological development, it became possible for authorities to stand upon power rather than upon sanctity, they did not dispense entirely with sanctity. Rather the relationship between sanctity and authority changed. Whereas the unquestionable status of ultimate sacred propositions previously rested upon affirmation through religious experiences of the faithful, it now came to rest, overtly or covertly, upon force. Whereas previously authority was contingent upon its sanctification, sanctity now became the instrument of authority. Coercion is difficult and expensive, and compliance and docility are achieved more easily and inexpensively through first the encouragement of religious experiences inspired by hopes of salvation in another life and, second, inculcation of the belief that the world's evils are a result of the worshipper's own sinfulness rather than a matter of external exploitation or oppression which the worshipper could possibly resist. (Rappaport 1971:41).

Rappaport's ideas are of a general evolutionary sort and suggest that religion will have less material use and fewer ties with subsistence the more effective technology becomes. This may be true in general, but it is also true for American utopias in 19th century industrialized society. As a consequence of having chosen isolation many placed themselves on the physical frontiers of the United States, and in the case of the Mormons after their move to the Great Basin in 1847, well beyond the frontier. This and poverty effectively removed the group from the sources of technology and technological innovation, and simultaneously had the effect of imposing what we can call technological primitiveness on a community which was also seeking both a simpler existence and freedom from industrial society.

Physical isolation often acted to reduce the level of technological sophistication in many utopias. In the face of such relative technical primitiveness any group would have been able to depend more heavily upon sanctity to govern key subsistence variables. While most utopian communities did not seem to be aware of the tie Rappaport suggests between subsistence and sanctity, many established a close tie between these two factors by breaking the hold of industrialization.

In most utopias physical isolation was only one factor that made sanctity preeminent over coercive power. Often complementing it was their notion of institutional completeness.

> . . . for the great majority of nineteenth century communes were more than simple agricultural societies; they were to some extent concerned not only about production of enough goods and services for the community but also about their commercial and political relations with the larger society. Most of them had businesses of one kind or another, sometimes servicing the community primarily but also exchanging goods with the outside society. Zoar (a German separatist utopia in Ohio, 1817–1898) had both agriculture and industry, including woolen, linen and flour mills, a timber planing business, and a wagon shop. Oneida (founded by John Humphrey Noyes, Vermont and New York, 1848–1881) in its early years engaged in farming and silk-jobbing besides running a flour mill, saw mill, and machine shop. It began successful fruit canning, bag-manufacturing, and animal trap businesses, and still later a successful silverware factory. Saint Nazianz (a Catholic commune in Wisconsin 1854–1896) sold cheese, beer, straw hats, shoes, and wheat. Among Harmony's (a German separatist utopia in Pennsylvania and Indiana, 1804–1904) items of commerce were hides, grains, furs, waxes, linen, tobacco, and cheese. The Shakers are still known for the furniture they manufactured. (Kanter 1972:150).

Institutional completeness avoided external dependence and was facilitated by isolation and in turn facilitated practical self-sufficiency. However, the practical self-sufficiency often led to prosperity and a yet deeper involvement with the outside world, as well as to the adoption by some utopias of production practices which reasserted the primacy of those very industrial practices utopias had been established to eliminate. As well as salespeople and customer relations specialists which several communities like Amana, Harmony, and Oneida maintained, the last also hired 200 workers at peak production times. Hiring outside labor was also a practice among the Shakers and at Zoar (Kanter 1972:151–2). Inherent in the successful marketing of products was a continual push toward organizational efficiency and away from practices maintaining community solidarity.

Utopias which had industries, especially successful ones, were caught between keeping the industry subordinate to the community's ideology and allowing a successful industry to become a capitalist enterprise. The struggle usually meant that the industries failed to compete with the world after a while because the community was deliberately using a technology more primitive than that used in the surrounding society for the same tasks, and did so to maintain the tie between its beliefs and the way it wanted to conduct daily affairs. Technological primitivism maintained the tie between ideology and subsistence, and this ultimately had the same effect that spatial isolation had on those communities which were so far beyond urban civilization that they were all but stripped of modern technology.

What we have seen so far is the use of isolation and relatively primitive technologies among American 19th century utopias. These same communities differed from American society mainly in that they sought to reshape the life of their members according to a set of religious ideals centering around, ". . . perfectability, order, brotherhood, merging of mind and body, experimentation, the community's uniqueness . . . harmony with nature, harmony among people, and harmony between the spirit and the flesh." (Kanter 1972:54). While these represent the conscious ideals of most of these groups, few of them were explicitly aware that realizing them in action had to involve simplifying subsistence practices to the point where the community had to rely on sanctity, the power encoded in the divinely inspired ideals mentioned above, to regulate key economic variables. In order to operate according to their principles most groups changed their technologies and their economies in the direction of simplification. This almost inevitably meant a more primitive technology in the sense that they lost some real control over factors central to the group's survival. As a consequence of giving up real control over aspects of subsistence, these groups automatically reverted to power derived from the supernatural to regulate what they no longer had the technological capacity to handle.

To use Rappaport's hypothesis: these groups chose to regress in an evolutionary sense. In fostering technological primitivism they moved back to a condition where sanctity was the only source of power available to govern key aspects of subsistence. And while this may be true analytically, the refounded tie between sanctity and subsistence is what utopian communities did deliberately seek, and it is, at least for a short time, what they got.

Having talked about isolation and its corollary, institutional completeness, in the context of the technological peculiarities of utopias, it is appropriate to turn now to sacred technology, one of the two more easily identified traits common to utopias. Given the amount of thought and planning they spent on the material conditions they wanted to foster, the existence of sacred technology is more easily understood. We can understand how ordinary items like dress and furniture could be used by a group of people to establish and maintain its own distinctiveness and how these items obtained a level of meaning that their analogues simply did not possess in the surrounding society. Maintaining deliberate and distinct differences is one way sacred technology comes into being. And it is also established when utilitarian tasks which are regulated by religious rules—when plowing, hat making, gathering hay, turning a lathe, churning butter, or turning out pig iron—mean preparing the garden for the Second Coming of Christ or are acts of worship. Such a level of meaning exists when technological primitivism has been established and when, as a result, sanctity is used to regulate the economic variables the technology is directly tied to.

Following are two examples of sacred technology and sacred work which technological primitivism and isolation foster. The first is also an example of how sanctity's power, which exists because of isolation and a primitive technology, regulates subsistence. Mormons in the Great Basin West employed irrigation agriculture which depended on systems of canals leading water from mountain-fed streams to the fields (Leone 1974:722–766). Central to the system was a diversion dam which created a raised pool of water that, having been created behind the dam, slowed the water and allowed its silt to settle. When the pond was high enough, the water flowed through the canal leading from the stream's bank. Because the river water was extremely silty, the ponds behind the low dams were quickly filled with solid material which made the pools behind them even shallower. This forced river water into the canals before the silt could settle out and consequently deposited silt along the canals and in the fields as the water slowed down. When a shallow pond failed to allow enough time for water to settle before it entered the canal system, canals automatically had to be cleaned more frequently and fields turned over more often to break the suffocating crust which silt forms.

The periodic floods which characterize much of this semi-arid region were rarely contained by a dam, and never by one which was silted up. They swept over a dam, eroding its foundations on the far side and finally collapsing it. The collapse allowed the flood to erode the collected silt and also directed the water down the stream bed, keeping the flood from washing out the canals and over the fields. The Mormon dams were primitive throughout the 1880s and 1890s because of a lack of adequate building materials, trained engineers, or technicians of any kind, and machinery other than iron implements and horsepower. Nonetheless, the Mormons invented collapsible dams which could be quickly built by amateurs, quickly washed away by a threatening and cleansing flood, and quickly rebuilt after a flood. Given the circumstances, these dams were of such tremendous utility that at Joseph City, Arizona eight were built and rebuilt and at Woodruff, Arizona thirteen.

When a dam went out in a flood, it was usually rebuilt as a cooperative effort by all the Mormons in all the surrounding towns, many of whom also depended on collapsible dams for their subsistence and who would perhaps need similar help. The whole rebuilding enterprise was tied to the church's ritual cycle. The funding, manpower, planning, and organization, the religious explanation for the collapse as a trial to the Saints, the equally religious call to meet the Lord's latest test in rebuilding the dam, were all provided through Mormon rituals. In Sunday Sacrament meetings, in the Quarterly Stake Conferences (regional meetings), in the many weekly meetings of the priesthood quorums encompassing all males, Mormons prayed and sang, worshipped and talked, preached and reasoned with each other about dam rebuilding. All this took place in the same meetings. After organizing their new effort in these rituals, they then went to the dam site, consecrated it and every new day's work, were directed by the priestly leaders in the project, and understood consciously how the rocks of the dam and the water raising behind it were an explicit and essential part of building God's Kingdom on earth. Ritual regulated the entire rebuilding cycle from the moment after a washout when disaster relief was necessary, to the regulation of water fees, land rights, dam construction, and the apportionment of crops grown in the irrigated fields after the dam was built. The 19th century records are unambiguous about this and show sanctity as the major variable regulating balance in the subsistence system.

By the 1920s, all over the Great Basin, including Arizona, the whole irrigation system had been secularized. The U.S. Departments of the Interior and Agriculture introduced technical improvements and a sophisticated technology into the water control system and placed these improvements in the hands of secular authorities. This eliminated the Mormon ritual system as the key regulator in the system. A more modern technology and set of management techniques divorced religious ritual from that part of the subsistence system represented by irrigation. When the Federal government inadvertently eliminated technological primitiveness, Mormonism and sanctity stopped regulating ecological variables.

The degradation of sanctity's power occurred for several reasons. First, Mormons had, and still have, a long-standing notion of earthly progress, which means greater control of earthly environments, or what is sometimes termed a this-worldly orientation. This idea makes them readily receptive to technological improvements even though, at the same time, some Mormon leaders in the 19th century understood that some forms of industrialized technology, like mining and the railroads, would undermine church control of daily life. Second, because rituals were key devices for governing the economy, many technological improvements were introduced through them. Thus the machines and techniques which were to undermine ritual's regulatory capacity were not infrequently sanctified by the very rituals the machines soon undermined. Third, because the federal government was the supplier of most of these technical improvements and remained in control of them, it replaced the church as guarantor and possessor of the local Mormon economy. The federal government had spent much time, trouble, and money trying to harness the Mormon church late in the 19th century, and finally broke its tight control over the economy of the intermountain west by massive gift-giving, something called the paradox of the gift-laden conqueror (Murra 1962:710–728). When the area's technology had been modernized, Mormonism emerged as a church; it was no longer the theocratic state it had been. Sanctity's power to control the economy was dead, and so was the Mormon utopia.

What we see among Mormons is technology and subsistence controlled by sanctity when, because of willful isolation on the frontier, an adequate irrigation technology was unavailable to them. Here the relationship between isolation, primitive technology, and sanctity is evident.

When we look at the Shakers, we can see how this matrix can endow otherwise ordinary items with a level of religious meaning usually absent from their analogues in secular society; in other words, how sacred technology is created and how it is related to the three variables cited above: isolation, primitive technology, and sanctity.

> Suppose a table were required for the ministry's order. . . The piece had to meet the minimum specifications for strength and convenience. But even a crude table would meet such standards. The Shaker joiner went further. The aim was to improve, to perfect, to make a table superior to any world's table. The 'Holy Laws of Zion' had admonished him to labor until he

> 'brought his spirits to feel satisfied': whatever is fashioned, they ruled 'let it be plain and simple, and of the good and substantial quality which becomes your calling and profession, unembellished by any superfluities, which add nothing to its goodness and durability.' 'Think not that ye can keep the laws of Zion while blending with the forms and fashions of the children of the unclean!' For conscience sake, then, the table must be well-proportioned, have grace and purity of line, as well as plainness akin to the humility of Shaker life. (Andrews 1950:336–7)

The process of furniture manufacture was an act of worship. The end product had a certain value and consciously represented the key precepts of Shaker life. The finished object's special meaning as distinct from similar objects in the surrounding culture is clear; nonetheless, the objects in themselves are neither sacred nor are they venerated. In fact, the key difference between these objects and those of the secular culture was that the end product, for Shakers, was far less important than the dynamic process of creating the piece. This process was the on-going act of worship, and this is what sacred technology looks like.[1]

To summarize this presentation so far, I have explored how isolation fosters primitive technology; how, under this condition, sanctity regulates subsistence; and how some technological items take on religious meaning under these circumstances. Next I shall explore the implications for utopias in choosing primitive technology to achieve success.

Given primitive technology as a feature of American utopian groups, there are two corollary principles we can apply. The first and more obvious is that when the parent society that gives birth to the utopian group no longer has a use for it or is threatened by its success, it can introduce its own more sophisticated technology and techniques into the utopia and undermine the key tie between religion and subsistence. Federal improvement of Mormon irrigation technology illustrated how this corollary works. And the general utility of the corollary is especially highlighted when we examine the wide range of experiments these utopias carry on for the parent society. In emphasizing the nature and extent of the contributions which utopian experimentation makes to the larger society, the importance of this corollary as a control mechanism will become clear. It highlights one of the automatic and peaceful ways an effective but independent utopian solution is reintegrated into the parent society whose problems gave rise to the utopia and its experimental solution in the first place. This next section allows us to see the other half of the reciprocating process between utopias and their parent society: the return contribution utopias make and how the contribution is managed.

In general we can say that most industrialized societies produce utopias, even though their advent in the Western world precedes industrialism. Most scholars have regarded utopias in the United States as functioning as laboratories for social experimentation, as homes for social deviants, and in general as a safe way of handling discontent. This hypothesis, which is one of the basic sociological views of utopian movements, sees them as escape hatches and safety valves for society's malcontents. It regards these new communities in two ways. Most often, it focuses upon the way a new community is organized and run; it views the utopia from the inside and examines the problem of founding a new community and securing the commitment of new members (Kanter 1972). Second, this view sees utopias as populated by deviants who, for the safety of the parent society, are housed while they play out their odd ideas and work out a new relationship with the parent society; it sees utopias as devices for reenculturation (Prince 1974).

This perspective is not so much wrong, as it is neglectful of the positive functions of utopias. We have understood for some time that utopias engage in many forms of experimentation, and in her survey of American utopias Rosabeth Kanter makes it quite clear that those of the 17th and 18th centuries were primarily concerned with religious innovation; 19th century utopias with innovations in social organization; and today's with psychological and social-psychological change. These represent the key domains in which novelty has been tried. Any of these groups was operating on a frontier, experimenting in vital zones with rearrangements of traditional human relations to solve a problem the parent society was facing. In addition to these vital zones, a utopia may also be on a spatial frontier, a vital zone of another kind.[2] When the role of the Mormons in settling the whole Great Basin for the United States is examined, the positive value of utopian withdrawal to the edge of society for the purpose of experimentation becomes obvious. Utopias often solve critical frontier problems for society; they may populate, develop, and secure whole frontier areas or they may collect thousands of alienated young people and minorities and resocialize them with "acceptable" values, funneling them back into mainline society as many communes do today (Prince 1974; Anthony and Robbins 1974). They perform these services long before society is equipped technologically (railroads for settling the West) or legally and socially (laws on drug use, community health-care centers of various types) to do the same tasks for itself. American society uses freedom of religious experimentation as guaranteed in the First Amendment to allow utopian groups sufficient latitude to do these jobs earlier.

Frontier status in the sense of isolation is sought out to allow freedom to experiment. But given the pariah status of many utopian groups, it is clear that the isolation of frontiers is also imposed upon them. Being a pariah is usually derived from the novel religious, social, and/or psychological experiments which the utopias engage in. These are often accompanied by pressures forcing groups into various wildernesses which, once occupied, give them both the freedom to continue experimenting and also the need to expand on their innovations. Couple this with the emphasis on sanctity and add a handicapped technology, and from it emerges a complex pattern of frontier expansion which, regardless of the type of frontier, involves the entry of religious groups into problem zones to try experiments impossible at home. Once successful, such experiments are meant to draw—and actually do attract—attention from the parent which then proceeds to reabsorb the group. The reabsorption integrates the successful experiment and simultaneously eliminates a competitive threat to itself.

Since utopias are a function of the larger society, insofar as they are based on technological primitivism, their dissolution is also a function of the larger society. The dissolution is sometimes violent, sometimes not. The Mormon and Oneida cases involved real confrontations with government; the Shakers did not. When there was confrontation, at least in the Mormon case, real control of the Mormon church and its big and quasi-independent population was achieved not through threats from Congress, the courts, or the army, although all were tried, but through massive gift-giving. This corollary says that America has a built-in way of destroying the very utopias it spawns. Industrial society undermines the technological primitivism which is needed to make a utopia work by giving it a more sophisticated technology, the fragmenting effects of which the utopia rejected in the first place.

The second corollary to the use of primitive technology by American utopian groups, is that American utopias are self-eviscerating. While many rejected industrialism, all looked for immediate earthly betterment, often called perfectibility, in this life (Kanter 1972:32–39). Mormons, Shakers, and all the rest understood the dangers of the factory system, railroads, mining, and the fragmentation of city and tenement life. And they all sought to create conditions which avoided these and which often resulted in fact, although not necessarily in conscious principle, in technological primitivism. Because the results of a primitive technology were rarely if ever clearly understood, the conflicts between it and immediate earthly improvement were just as rarely or never seen. When, after the Civil War, Shakers began using mechanical lathes to make their furniture, they themselves say they saw their sect go into decline. When in the 1890s Mormons began to use improved farming techniques and had a surplus to market, they saw their own society enveloped by the same one they had rejected sixty years earlier. If either group had had a well-articulated notion of technological primitivism, they would have debated with themselves more seriously than they did before utilizing more convenient, more successful, and more powerful technologies. But what they did have was a well-articulated notion of earthly progress which made them receptive to the very machinery and sophisticated techniques that ultimately made their dependence on the power of sanctity unnecessary. In other words, even if Mormonism had never been attacked by the federal government, it would have been as transformed as the Shakers were when they began using machinery to make their chief product. The Mormon theocracy was bound to collapse, indeed it already had begun to weaken, regardless of government intervention. In being tied to earthly success it was tied to a principle which was inevitably bound to separate religion, or a source of power which is based on the unempirical, from control over economics and ecology. With this corollary we can see the close systemic relationship between utopias, which are attempts at reforms in a capitalist industrial society, and the use of the reforms to strengthen, expand, and stabilize the parent society but without major political risks. This technique contrasts strongly with the analysis of the same problems in industrial society offered by Marx who suggested class analysis and revolution to achieve them. What we seem to have built in this country are religious, social, and psychological experiments which, when they have finished their task, or are too successful at it, either wipe themselves out or are fairly easily undermined from the outside.

As a consequence of this analysis we can see the function of utopias in this country as conservative, not as radical or destructive. They are a device for reforming and for housing reformers while the experiment and the experimenters work safely with changes the parent society always has the power to accept or eliminate. They are a key and necessary part of our society's expansion; they are vital to America and it is vital to them that they remain different from it.

This section has dealt with how utopias are controlled once their experiments are finished or get out of hand. It suggests that utopias self-destruct, but that while alive they often achieve tasks for the larger society that it is incapable of doing for itself. It is likely that the kinds of experiments utopias try can occur only under conditions like frontier isolation and religious rule. In other words, what society needs accomplished and what utopias need to exist may be identical. Society needs frontiers populated and developed and the alienated reenculturated and made productive. Utopias need to break the traditional controls of technology, need to experiment with new social and ideological concepts, and need the freedom provided by isolation to work out their novel changes. These mirror-image needs allow us to see how utopias and the society they spring from operate in a reciprocal relationship; they solve problems both recognize as real, even though in any one case the problem the utopia concentrated on and the one it solved for the parent society might not be the same. The Mormons did not set out to claim the Great Basin for the United States, they set out to found a sacred communist kingdom.

This view allows us to see why utopias are necessarily short-lived. Their evanescence has always been a problem both to themselves and to scholars, but when we see their positive problem-solving function we can also see that once a utopia has solved a problem, spatial, social, religious, or psychological, it can be reabsorbed or circumvented. Since most are based on sanctity, when the region they settled in is no longer on a frontier, the technological base for sanctity's operation evaporates and so does the tie between ideology and daily life. The frontier utopia is merely reabsorbed. It settled an area and even though that may not have been its own goal vis-à-vis the nation it came from, it had accomplished that task. Social and psychological frontiers last as long as specific social ills are unremedied and until the processes producing an alienated subpopulation stop naturally or are corrected. When solutions like social legislation governing working conditions, or today's drug care programs and community mental health care centers begin to work, then the grounds for the utopian rebellion are eliminated, and their existence is compromised, their populations are limited, and their solutions encompassed. In short, utopian life is limited both because of the technological basis for their religious sanctions and because their existence is so completely a reflection of the needs of the parent society. Hence, the length of their lives is a function of the time the greater society takes to invent its own solutions to the problems the utopia is solving, or the time it takes to absorb the solution the utopia itself offers. It is important to remember here that utopias are not ethnic groups with already established identities including histories, nor are they coherent minorities seeking independence in the sense of nationhood. Utopias and communes start off as incoherent groups of people whose identity exists only in opposition to the parent society and its problems. Utopias consequently have and seek no real independence, and as a result, when the major society redefines itself, it also redefines its utopias costing them their reason for being in the process.

The contributions made by American utopias are often not to be taken as the most obvious products we associate with them. Their major contributions may be settling frontiers or reenculturating alienated generations, but both the utopian groups themselves and popular opinion tend to suppose that Shaker furniture, Oneida silverware and the Graham cracker (the supposedly perfect food invented by a 19th century utopia) are their most noticeable gifts. These gifts are indeed noticeable for their uniqueness which is itself derived not so much from the objects' function but from their style, and style in turn will tell a great deal about the group's relation to its parent society.

Utopias experimented with many of the material items in the built environment they had to deal with. Almost everyone practiced a form of environmental determinism (Leone 1973:125–150), and all believed that town plans and architecture as well as the basic form of work, factory or farm, influenced behavior and could be manipulated and designed to produce the social results the community wanted to foster. House plans, barn, factory and town plans, furniture, clothing, food, and transportation were all consciously redesigned by one or another American utopian group. Some of these experiments were quite successful and produced items that enabled the group to control its own environment better than would have been the case with the analogous item from contemporary society. Further, in the redesigning process many of these pragmatic and utilitarian forms were given a level of meaning which showed that the item also operated in the religious or ideological realm; it is these that fall under the label sacred technology. We can expect that a Mormon town plan, Shaker furniture, and Oneida silverware all show traces of what their respective groups were accomplishing.

The particular style of these items reflects the conscious ideas that the groups wanted to express as well as the less conscious cultural processes, like isolation and rebellion, which are behind utopias in general. Being isolated and in rebellion does not mean complete severing of ties; quite the reverse. These groups all regarded themselves as guides and beacons to America. Consequently being both American and un-American, even anti-American, created a constant tension underlying them all, and the style of their material culture reflects both their uniqueness and the deliberate Americanness of that uniqueness.[3] As a result we can see that the style of utopian technology reflects not only the isolation of frontier conditions, the attempt to handle hitherto foreign environments, and a deliberate avoidance of technologies which routinize and brutalize work, but also that it reflects the constrained rebellion and separation which motivate all these concrete efforts.

Here it is important to explore what style is. Style is most obviously imposed on form, forms like architecture, sculpture, urban plans, painting, ceramics, the minor arts. Those are material forms. It is also imposed on the forms we know as music, prose and poetry, and dance. The Baroque is a good and common example of a style most scholars would agree is found in all these forms. When style is imposed on such forms, it is separate from their content, although obviously it shapes how the content or explicit message is expressed. I think we can also expect that in a single culture like that of the 19th century Mormons or Shakers, whatever the style is, it will be found in all these forms. If that is so, then the forms merely contain and express style, but they are not where we should look for its source. Style may be a manifestation of a more or less consciously articulated idea formulated by a society's intellectuals and found as a prominent symbol in its ideology, but it is also the result of material factors and a manifestation of some basic economic or political relationships found in the society.[4]

Consider the forms used by Mormons in the 19th century which everybody agrees had a particular shape. In the Utah period Mormon town planning produced uniformly distinctive towns all over the Great Basin. The landscape around the towns, the town grid, the uniform barn yards, the houses, many of which had similar floor plans, all are readily separable from what is found in all other western towns. Beyond this there is distinctive Mormon music, dress, a literary tradition beginning with the Book of Mormon and including an unbelievably rich out-pouring of journals, novels, plays, poetry, and history. Here is a huge number of forms. What unifies these expressions?

I cannot give a complete answer to that question. There are, in fact, forms like music I can't really address at all. But, assuming that they all are unified, I think the principle which creates the unity is mutually exclusive compartmentalization: the use of categories whose closest members have no contact with each other.

This principle which characterizes 19th century Mormonism (and which may characterize modern Mormonism as well) was found in Mormon town plans, farm yards, and house floor-plans. In the 19th century (and even today in the Southwest) a Mormon family's holdings in town or in the countryside near a town were composed of a series of yards, pens, and enclosures all in a very small area and all functioning to provide a protective micro-environment for a plant or animal species which, if not closely isolated in this semi-arid environment, would not survive productively. The barnyard environments acted to separate predator and prey, and as such they remained separate and closed to each other. They had to have close mutual borders, but no contact or mixing was allowed. Here was real ecological crowding, very close living forced by a desert environment and lack of water. It was successful because it was essential, but incompatible categories were kept strictly isolated from one another. Fences, streets, hedgerows, and long lines of trees created and marked these borders and also formed the visible identity of a Mormon town in the West; they still do and, of course, this is one of the most apparent stylistic differences between Mormon and non-Mormon towns. To a Mormon, all this singularly ordered material culture identified redeemed Zion from the unredeemed. But beyond an expression in form, the redemption could be brought about only by operationalizing the general principle of closely juxtaposed but exclusive compartments; otherwise subsistence would never have been a reality.

Nineteenth century Mormons were often polygamous, and when a man took a plural wife a religious injunction specified completely equal treatment of each new wife. This meant that a man either had to set up a house on a separate and equal plot of land for each spouse or, as was more often the case, provide equal apartments or bedrooms in one house for each wife. The result was the isolated separate apartments, the multiple doorways, staircases, and gates which are an often-pointed-out stylistic feature of Mormon towns in the Great Basin. All wives were equally important, and had to be treated not as a collectivity but as individual women. A man did not have one large family, he had as many equal families as he had wives. They were not necessarily to be kept physically separate from each other, in fact there was often great cooperation among them and warm emotional bonds as well, but these could develop because of the concrete strictures on tight conceptual and material separation among the equal family units. Here we can see, both in social organization and in architecture, the separate compartments which can live together successfully only because they are not allowed to impinge on each other's domains. The use of categories whose closest members have no contact with each other accounts for the peculiar look of a Mormon yard and of a Mormon house. It also operates through Mormon social organization. This observation leads us to explore the source of the principle which is itself the source of style.

In looking first at Mormon ideology, especially the consciously articulated part of it, there is an idea which fits the stylistic principle more or less precisely. "Opposites must exist—good and evil, virtue and vice, right and wrong—that is, there must be an opposition, one force pulling one way and another pulling the other. Agency (a Mormon idea which is built on the notion of free will) is the philosophy of opposites, and because these opposites exist, men can reap either salvation or damnation by the use they make of their agency." (McConkie 1966:26–27). This idea speaks of the need to choose and makes it plain that choices are a part of living, that the world is full of acceptable and unacceptable compartments called good and evil. Mormon doctrine decries predestination and in doing so replaces it with a categorized universe whose compartments do not allow for areas of grey but create a continual tension between corruption and incorruption, sense and insensibility, happiness and misery. The doctrine cites the need for such choices, it does not specify where the choices are to be encountered, nor how to make them when found. In specifying a world of choices it consequently brings compartments into existence.

The doctrine sets up life as a series of incommensurable categories, or at least of paradoxes to be dealt with. It certainly is one source of close but mutually exclusive categories.[5] Nonetheless, the content of those categories and the choices which had to be made among them come from early Mormon history.

The whole doctrinal idea describes the paradox of Mormon existence before entry into Utah. And in general it describes Mormon life within but always actively apart from America. Mormons always wanted to be and to remain American, but they also and more vocally wanted to be different, separate, distinct, and to be an improvement on and better than America. They wanted to be distant and close at the same time. Ideologically Mormons held the United States Constitution to be divinely inspired and regarded the United States as an unusually favored nation. But they also anticipated the quick downfall of earthly governments which were all hopelessly corrupt, including the American. Their own priesthood would then hold earthly power until the Second Coming made such a burden unnecessary.

This intellectual or ideological ambivalence lent itself to a program of action in which Mormons set themselves up three or four times in the 19th century as a separate community within the borders of the settled United States. Each time they did so and enjoyed a visible measure of material success which stemmed from their efficiency and hard work, they proclaimed their success as coming from those policies which were least like mainline America. Their seeming arrogance and their obvious success led to expulsion from New York, Ohio, Missouri, and Illinois. Even in Utah they had to fight hard for survival. When in 1890 the Mormons had been thoroughly pommeled by the federal government and even seriously threatened with loss of U.S. citizenship, the most they could manage to feel was that their rights as United States citizens had not been respected. Mormons knew they were different, wanted it that way, and were successful because of those differences. They were not so aware of how American they had remained until the power of the government and the emotional force of losing their American identity faced them. With all these confrontations, Mormons managed survival and managed to be Mormon and American at once. Mormon 19th century history is a dialectic between their two goals: remaining American and being distinctive. And their culture, including their material culture, is a result of that mutual interaction. Mormons have had to invent a system that holds two incompatible entities at once: deifying America but condemning its government; being theocratic, polygamous, and dissolving private property while demanding freedom of religion, equality before the law, and statehood for Utah. This system of paradoxes has ideological roots, has roots in economics and politics, and is expressed without much conscious intent in many, and I would not be surprised if it turned out to be in all forms Mormons utilize to build and express their own environment. The contradictions that are part of the reality Mormons have to live with are expressed throughout their culture in a way we may recognize as style. This relationship is probably true for every American utopia.

Utopian styles are a manifestation of the variables explored in the first part of the paper: isolation, primitive technology, and the use of sanctity to run the economy. They are a manifestation of being un-American in America; they simultaneously are a product of that tension and an aid in resolving it.

This discussion of style has really been a discussion of the oppositions or paradoxes within Mormon culture. From that viewpoint Mormonism has been seen as a structuralist would look at it: a single set of opposites played out through all the areas of Mormon culture allowing us to see it as a coherent, unified whole. Here I have obviously chosen to call that unity style. The style itself is the product of Mormon history, and that history is the creation of a unified but distinct group within the borders of the United States. Mormonism, like Shakerism and the other utopias, is readily visible because it was distinct and systematically different. And its distinction and systematic differentiation, that is, its structural make-up, stem from the dialectical processes that gave birth to 19th century utopias in the first place. They were in rebellion and sought the isolation of frontiers. But the rebellion was meant as a vanguard action, and the frontier was the parent society's. Utopias were simultaneously in and out of America, and internally were never completely separate, integrated wholes. They never desired total independence from America, and as a result of their internal structure—unresolved contradictions and the consequent tension—they enjoyed only brief lives. Because they never separated themselves they died young; and the inability to separate was born of technological primitivism—rebellion against industrialism—which undercut their independence from the beginning. To come full circle, we see the structural tension within utopias which was produced both by their rebellion and by the actual method through which they realized that rebellion: partial but never total separation; isolation, but never with self-sufficiency; differentiation, but never with integrity (wholeness). This produced their style. Structure and style as outlined here are in turn greatly influenced by material factors such as the total social disruption caused by the Industrial Revolution, the existence of spatial frontiers, and the pull between these two as the industrializing world looked at frontiers as zones for potential exploitation and the already exploited looked to the same areas as zones to escape further disruption. The pull between the two within 19th century America led to the birth of our utopias which we can now see are simultaneously products of that material process, and endowed by that process with an internal structure that guaranteed they would participate in it and not free themselves from it.

I began this look at 19th century American utopias with a materialist analysis, proceeding then to a structuralist one, and linking the two by referring, ultimately, to a structure which illuminates isolation, frontiers, and rebellion—those issues that were raised at the outset of this discussion. In doing so, I hope I have highlighted a way of using materialism and structuralism which assumes a noncausal or systemic relationship between the two.

NOTES

[1]I am grateful to Donald Smalley for providing me his research on Shaker material culture.

[2]The positive function of utopian experimentation in the larger society was suggested and outlined for me by Robert Schuyler of the Department of Anthropology, City College, CUNY.

[3]Emory Elliott of the English Department, Princeton University, pointed out the tension within utopias which is the source of style.

[4]This discussion of style evolved in several conversations with James Boon of the Department of Anthropology, Duke Unversity.

[5]I am grateful to John Sorenson of the Department of Anthropology, Brigham Young University, for pointing out the Mormon idea of "opposition in all things" to me.

LITERATURE CITED

Andrews, Edward Deming and Faith Andrews
1950 Shaker Furniture; the Craftsmanship of an American Communal Sect. New York: Dover Publications.

Anthony, Dick and Thomas Robbins
1974 The Meher Baba Movement: Its Effect on Post-Adolescent Social Alienation. *In* Religious Movements in Contemporary America. I. I. Zaretsky and M. P. Leone, eds. Pp. 479–511. Princeton, N.J.: Princeton University Press.

Kanter, Rosabeth M.
1972 Commitment and Community. Cambridge, Mass.: Harvard University Press.

Leone, Mark P.
1974 The Economic Basis for the Evolution of Mormon Religion, *In* Religious Movements in Contemporary America. I. I. Zaretsky and M. P. Leone, eds. Pp. 722–766. Princeton, N.J.: Princeton University Press.

1973 Archaeology as the Science of Technology. *In* Research and Theory in Current Archaeology. C. L. Redman, ed. Pp. 125–150. New York, Wiley-Interscience: John Wiley and Sons.

McConkie, Bruce R.
1966 Mormon Doctrine. Salt Lake City: Deseret Press.

Murra, John V.
1962 Cloth and Its Functions in the Inca State. American Anthropologist 64:710–728.

Prince, Raymond H.
1974 Cocoon Work: An Interpretation of the Concern of Contemporary Youth with the Mystical. *In* Religious Movements in Contemporary America. I. I. Zaretsky and M. P. Leone, eds. Pp. 255–271. Princeton, N.J.: Princeton University Press.

Rappaport, Roy A.
1971 The Sacred in Human Evolution. Annual Review of Ecology and Systematics 2:23–44.

*

DIFFERENT APPROACHES TO TECHNOLOGY

5

Technical Skills and Fishing Success in the Maine Lobster Industry

JAMES M. ACHESON

University of Maine, Orono, Maine

I. INTRODUCTION

Anthropologists working in hunting and fishing societies around the world have long noted that some individuals have a great deal more success than others. This is true not only in primitive hunting-gathering societies,[1] but in modern fishing communities as well. In the literature on American and Canadian fishermen, reference is often made to "high-liners," "good" fishermen, "bad" fishermen, and so on (Acheson 1975b; Proskie 1971:52). While the reasons for differential success have not been explored, there is increasing evidence that the technical skills of the captain are far more critical in determining the catch of a boat than any other single factor.[2] However, in the whole anthropological literature on fishing communities there are very few studies concerning skills, and these are not comprehensive by any means (e.g. Davenport 1960:3–11; Harris 1968:581; Wadel 1972:111). There is no literature on the skills of New England fishermen. This oversight is not surprising, given the fact that anthropologists, sociologists, and others, have made few serious efforts to investigate any aspect of technology.

However, economists, engineers, and others interested in technology have long noted the importance of skills. (In fact, the concept of human capital—the study of the effect of education and skills—has become a very important one in economics [Becker 1964]). For people in these disciplines it would be a matter of common sense that skill has an effect on output, albeit one that is difficult to measure. This paper attempts to go beyond the obvious to make three specific arguments about technical skills in the Maine lobster industry.

Argument 1: Technical skills play a very important role in determining output in the Maine lobster industry relative to physical inputs. In any industry, production is influenced by several inputs. The problem is to separate out the effects of these various factors. In section II, the results of our regression analysis are presented, which indicate that output cannot be explained by reference to physical inputs alone. The large residual strongly suggests that skills play an important role.

Argument 2: Skills in trap placement are more important than any other skills. According to the fishermen themselves, success in lobstering is largely due to the ability to place traps to maximize catch relative to trap losses. This necessitates a knowledge of a wide variety of natural phenomena. In section IV, these skills are identified and discussed in light of information from fisheries biology.

Argument 3: The skills fishermen identified as "critical" have a measurably significant effect on the output of individual fishermen. In section V, these skills and output are linked together by a controlled comparison of thirty-three fishermen who have different levels of skills. These data indicate that men who know how to place traps on "good" bottom with pinpoint accuracy earn higher incomes, catch more lobsters, and catch more lobsters per unit of effort than "unskilled" lobstermen.

II. ARGUMENT 1: THE RELATIVE IMPORTANCE OF TECHNICAL SKILLS

The evidence indicating that technical skills are critically important in the Maine lobster fishing industry comes from a variety of sources. Maine fishermen are very quick to say that "some men know how to catch fish and others cannot." I know two men who fish out of the same harbor. The younger, who had only five years experience in lobstering in 1973, used 900 traps and a thrity-six foot boat. The older man, with some thirty-six years of experience in the business, used only 300 traps, which he fished from a boat twenty-eight feet long. In spite of these differences in equipment, the older man had a net income of $21,000.00 that year from lobstering while the younger man netted only $12,800.00. Moreover, men who go out of the fishing business will often say that their failure was due to the fact that they did not really know how to use their equipment to full advantage[3].

The difference between "good" and "bad" fishermen is clearly more than a matter of capital equipment or effort. Dealers and fish buyers affirm that some men in the industry earn vastly more than others using the same equipment in the same area. However, if technical skills are important, they are very elusive, and study of them does not easily lend itself to statistical analysis. It is not only that fishing skills are intangible; the problem is compounded by the fact that skills are only one of a set of variables influencing output.

A fisherman can combine skills and other productive factors in a variety of different ways. Skills in lobstering can be traded off for capital (e.g. number of traps, size of the boat, horsepower of the engine), for hired labor, and for sheer effort. This means that there is no typical lobstering operation from an economic point of view. Some men who have big boats, large numbers of traps, and sternmen to assist them, catch no more lobsters over the course of a year than more skillful men employing smaller boats and using half the number of traps.

Although it is impossible to separate out skill from these other productive factors and directly measure its effect on output, there is some statistical evidence suggesting that fishing skills are a critical variable. In 1971, a research team from the University of Maine gathered a great deal of social and economic information on fishermen from three communities in Maine; a year later the questionnaire used in this study was given to a random sample of 126 lobster fishermen from other communiies.[4] A statistical analysis of these data revealed some surprising information about the factors influencing the production of lobsters.

In most industries, one can ordinarily find a close correlation between physical inputs and physical outputs. This was not the case here. An analysis of these data using both bivariate and multiple regressions revealed no significant relationships between income and number of traps used. When the number of trap days was run againt income earned, the R^2 was .55051 which indicates that we can explain only 55 percent of the variance in income by knowing the number of traps a man had in the water over the course of a year.[5,6] When income was run against total investment, the results were, if anything, worse. The R^2 was only.39439.[6] When income was run against the number of days fished, the results were no better. The R^2 was .21595, indicating that only 21 percent of the variance in gross income can be explained by knowing the number of days fished.[6] Expected income cannot be explained in terms of the ages of the fishermen either. When income was run against age, the R^2 was only .25767.[6] While the meaning of this last figure is far from clear, it is important to note that as men get older they tend to fish fewer days and fewer traps than they did when they were in their thirties. By the time they are in their sixties, they are usually partly retired, have sold most of their traps, and fish from an outboard-powered skiff. One might assume that since many older men are more skillful than younger men, age and income woud be highly correlated. These results indicate, as nothing else could, that age is not a good surrogate variable for skills.

The results of some of the multiple regressions point to the same conclusion. In one equation, expected gross income was assumed to be a function of trap days per year, days per year fished, education, months per year fished, and total investment. The R^2 for the total equation was .6154 which indicates that only 61 percent of the variance in income can be explained by all these factors in combination. (Applying a standard F test, the R square value is significant at the .01 level.) This leaves an unexplained variance of 39 percent. The fact that the residual is so large suggests either that a great deal of the variance is due to one or more variables which were not included in the equation or that chance plays a large role in influencing catch.

The equation for the multiple regression was:

$$Y_p = a + b_1X_1 + b_2X_2 + b_3X_3 \ldots . b_kX_k$$

Where

Y_p = expected income
X_1 = trap-days/year; b_1 equals .15360
X_2 = days/year fished; b_2 equals 49.4144
X_3 = education; b_3 equals 451.204
X_4 = months/year fished; b_4 equals 1028.59
X_5 = investment; b_5 equals .70749

A standard two-tailed test of significance was applied to all five variables. The t values of X_1, X_2, X_4, X_5, proved to be significant at the .01 level. The t value of X_3 was just slightly less than the critical value necessary to establish significance at the .05 level.

In cases of this kind, it has become commonplace for economists to assume that output is attributable in large part to technology and skill

(Scherer 1970:346–47). We would like to suggest that if some variable measuring skills had been included in the equation, the residual would have been far smaller.

While all this evidence is subject to qualifications, it does suggest that skills are important in determining catch. But it certainly proves nothing conclusive beyond the fact that the output of lobsters cannot be predicted merely by knowing the physical inputs.

In the final analysis, this preliminary study was of primary value in that it forced us to take seriously the fishermen's statements about the importance of skills, and stimulated an intensive study of the relationship between "critical skills" and the output of fishermen under controlled conditions. The results of this further study are reported in sections III, IV, and V.

III. GENERAL FEATURES OF THE MAINE LOBSTER INDUSTRY

In order to understand many facets of arguments two and three, it is necessary to have some background information on several diverse aspects of the lobster industry.

Technical Features

1. Equipment. The American lobster (*Homarus Americanus*) is found in the waters off the Atlantic coast of North America from Newfoundland to Virginia. However, Maine consistently produces far more lobsters than any other state.

The technology employed by lobstermen along the entire length of the Maine coast is relatively uniform. Lobsters are caught in wooden traps, or "pots," about three feet long, made of oak frames covered with hardwood lathes. Lathes are placed about 1.5 inches apart, allowing free circulation of sea water while retaining the large, legal-sized lobsters.[7] The open end of the trap is fitted with a funnel-shaped nylon net, or "head," which lets lobsters climb in easily, but makes it difficult for them to get out. The traps are attached to a small styrofoam buoy via a "warp" (polyethelene or hemp rope). The buoys belonging to each lobsterman are marked with distinctive sets of colors, registered with the state. These traps are baited with fish remnants obtained from nearby processing plants. The traps are usually placed in the water "in strings," or long rows, so that a man can see from one buoy to another in the fog.

Since World War II the relative costs of bait, fuel, lobstering gear, and boats have fallen relative to the cost of labor. The result is a marked trend toward a more capital intensive industry. Over the course of the last twenty-five years, men have been building larger boats and equipping them with much larger engines and cabins with windows. While these boats are far more expensive to build, and consume a great deal of fuel, they do allow a fisherman to carry more lobstering gear, and their stability and comfort make it more feasible to fish in bad weather and during the cold winter months. In addition, the large boats allow lobstermen to exploit a far larger area than before.

The most important innovation in the past thirty years is the hydraulic trap hauler which allows the number of traps operated in a day to be greatly increased. Before the early 1950s, when the hydraulic hauler came into general use, the average lobsterman could only "haul," empty, and rebait 125 traps a day by hand or with the old style winch; today, he can easily haul double that number using the newest kind of hauler.

There is no doubt that the hydraulic hauler has caused a dramatic increase in the number of traps being used. In 1953, when the hydraulic trap hauler had just come into general use, there were an estimated 440,000 traps in the water; in 1973 there were 1,747,000, almost a 400 percent increase in 16 years (Dow 1961:10).[8] Although fishing effort has increased dramatically, the number of full-time lobstermen has not greatly increased during the last ten years.[9]

Today most lobstermen fish alone from gasoline or diesel-powered boats 28 to 34 feet long, equipped with a depth sounder, hydraulic "pot" hauler, ship-to-shore radio, and compass. In the island areas boats may be somewhat larger, more often diesel-powered, and also equipped with radar to cope with the more violent offshore seas and the fog. In 1973 it cost between $13,000 and $20,000 to have a boat constructed. In addition, a lobsterman may have from $8,000 to $10,000 invested in traps and fishing equipment, a pickup truck, dock, and some kind of workshop. Replacement values easily run over $50,000.

2. Timing. A lobsterman's activities vary greatly from season to season. The midwinter months are unquestionably the slowest time of year. During January, February, and March, when men fish three to ten miles offshore, lobstering is generally more dangerous and unprofitable. Bad weather and high winds increase trap losses and make the work more difficult. Some men stay ashore during this period to build lobster traps, while others use their boats for scalloping or shrimping. Those who persist in lobstering during the winter may pull their traps no more than six or seven times a month. Spring (April 15 to June 15) and fall (August 15 to November 15) are unquestionably the busiest months of the year, when men have a maximum number of traps in the water and pull them every chance they get. During the three or four week molting season (June 15 to August 15, depending on the area) traps are typically placed very close to shore—literally feet away from breaking surf. During this period, catches are so small that men bring many of their traps ashore and do maintenance work on their boats. In the fall, lobstermen begin to move their equipment into deeper water again.

Throughout the year, lobstermen pull and rebait their traps when the sea is calm—usually in the early morning hours. Later in the day, when the sea is rougher, they have difficulty finding their buoys and operating their hydraulic trap haulers.

Institutional Features

1. Marketing. Any sizeable harbor has at least one lobster dealer who buys from local lobstermen and sells to tourists or to one of the three or four large wholesale firms distributing lobsters in Maine and the nation. Typically, a lobster fisherman maintains a long-standing relationship with only one dealer and sells his catch exclusively to that dealer. While the prices received by lobstermen do fluctuate seasonally, there is little price competition in the Maine lobster industry. However, it is important to note that "dinner lobsters" (at least 1.5 pounds with two claws) bring a higher price per pound than the smaller lobsters or "culls." Throughout most of the year, the price differential is about $.20 per pound at the Fulton Fish Market in New York. In many places this difference in price is passed on to the fishermen. As we shall see, this is one of the factors raising the income of "skilled" fishermen.

2. Territoriality. From the legal view, anyone who has a license can go lobster fishing anywhere. In reality, far more is required. To go lobster fishing at all, one needs to be accepted by the men fishing out of one harbor, and once one has gained admission to a "harbor gang," one is ordinarily allowed to go fishing only in the traditional territory of that harbor.[10] Interlopers are strongly sanctioned, sometimes verbally, but more often by the destruction of lobstering gear. This territorial system is entirely the result of political competition between groups of lobstermen. It contains no "legal" elements.

Violation of territorial boundaries meets with no set response. An older, well-established man from a large family might infringe upon the territorial rights of others almost indefinitely, whereas a new man or a "part-timer" would quickly lose a lot of fishing gear. Ordinarily, trap cutting involves only one or two men from competing areas. But perhaps once a decade, a series of small incidents will escalate into a full-fledged "lobster war," involving dozens of men and resulting in widespread destruction of lobstering gear. However, all conflicts are kept very quiet since trap cutting is illegal, and silence reduces the chances for a victim to retaliate. As a result, the public knows very little about the territorial system, or the political mechanisms that maintain it.

3. Prestige and fishing success. Maine lobstermen feel very ambivalent towards men who catch a lot of lobsters. On one hand, a great deal of prestige acrues to "highliners"—good fishermen who catch a lot of "fish" and earn high incomes. The most prestige goes to the men who let it be known in quiet ways that they earn a high income by skillfully working a small or moderate number of traps. Such a man is often elected to town office, his advice is sought by other fishermen, and he is very apt to serve as spokesman for the harbor gang in dealing with outsiders of all kinds. The prestige accorded such a man will increase as he gets older, but even a young man who is a "good fisherman" will be greatly admired and respected in a coastal town.

On the other hand, the prestige accorded a "highliner" may not completely negate the feeling that his success is at someone else's expense. Men who fish huge "gangs" of traps or who fish when the weather is bad are often considered to be taking advantage of others, indeed, to be "taking the food out of someone else's mouth." Such "pigs" or "hogs" can stir up a good deal of antagonism. Feeling against such a man may run particularly high if he is a braggart and his high income is due more to effort and capital equipment than skill.

This reaction against "success" stems from two sources. First, lobstering territories are considered the property of all the men in a particular harbor gang. There is a strong feeling that all the men in the "gang" should have an equal share of the lobster resource. Second, Maine lobstermen view fishing as a zero-sum game. They and fisheries biologists are convinced that there are only so many legal-sized lobsters in any given area of water. If one man catches them, obviously another cannot. This is not to suggest that men who catch few lobsters are highly regarded. Indeed, a "poor fisherman" with a lot of equipment is an object of ridicule. However, there are social pressures in harbor gangs which undoubtedly cause men to curb their fishing effort somewhat.

In essence, the rules defining success among lobstermen conflict. On the one hand, a man is motivated to compete with others to earn a high income and to demonstrate his success by conspicuously consuming that income according to the dictates of American society. Another set of norms defines a "good man" as an easy-going, helpful fellow who takes only his "share" of the lobsters. He is a skillful fisherman but one who does not escalate the level of competition.

Most fishermen attempt to escape from this double bind by being very secretive about the number of traps they have, their catches, and their income. Of course, other men can see where a man has traps, but they have no way of knowing how much he is catching from them. This is information that fishermen rarely talk about.

As a result, it is very difficult to learn the skills of lobster fishing. Skilled men will often give hints to their sons and nephews, but rarely to anyone else. Many lobstermen make it a point not to instruct their sternmen, who are potential competitors. Most lobstermen learn many of the important skills of their occupation by lone experimentation. Under these conditions, it is not surprising there are no standard terms regarding skills. On several occasions men described phenomena or techniques in such different terms that I was sure they were describing the same thing only after repeated interviews.

4. Heterogeneity in the Maine lobster industry. The lobster industry in Maine is not a homogeneous entity by any means. For our purposes, it is important to note three sources of variation in the industry. First, since there is so much variation in size of boats, propulsion units, and so on, it is very difficult to define the typical boat working under typical conditions. Lobstering is done with boats ranging from 16 foot, outboard-powered skiffs, to 45 foot, diesel-powered vessels with $10,000 worth of electronic gear and a three man crew.

Second, there are great differences in fishing practices in different parts of the coast. There are, for example, areal differences in numbers of traps fished. In the Casco Bay area some men fish as many as 2400 traps with a large boat and a two or three man crew. In this area it is customary to attach six to ten traps to one "warp"—a practice called "fishing trawls," as opposed to "fishing singles." On the rest of the coast it is not uncommon for men to fish 600 traps (singles or doubles), but the average man probably fishes 400 traps by himself with a boat no larger than thirty-five feet. In a good day, a man "fishing trawls" might easily pull 400 traps, while a man "fishing singles" will pull about 200.

The most important differences concern territoriality. In another publication, I have distinguished between *nucleated* areas and *perimeter-defended* areas and discussed their origin and their economic and biological effects (Acheson 1975a:193ff). The boundaries of *perimeter-defended* areas (on the outer islands) are sharply defined and are defended to a yard at all seasons. In these areas the sense of "ownership" extends up to the boundary line and perhaps a little beyond. The men fishing these areas feel that the whole area is theirs exclusively. They do not do any "mixed fishing" with men from any other harbor gangs. In the *nucleated* territories, an area within 2 or 3 kilometers from the harbor is usually maintained for the gang members exclusively, but the sense of ownership grows weaker as one goes further from the harbor. Far from the harbor mouths, there is a good deal of "mixed fishing," so that sometimes men from as many as five harbor gangs are fishing together.

While entry into any harbor gang is never easy, it is vastly more difficult to gain entry into harbor gangs which maintain *perimeter-defended* territories. This means that there are far more men per square mile of fishing area in *nucleated* areas than in *perimeter-defended* areas. In addition, men in *perimeter-defended* areas have self-imposed conservation measures (e.g. limits on the number of traps that can be used, fishing seasons, etc.). This difference in the rate of exploitation is clearly having differential economic and biological effects. The men in *perimeter-defended* areas are earning higher incomes than those in *nucleated* areas, and they are catching larger lobsters and more pounds of lobsters per trap. There is also indirect evidence that there is a larger breeding stock in *perimeter-defended* areas as well.

5. Controls. Since there are so many economic and institutional factors influencing catches, the role of skills can only be studied if strict controls are employed. In particular, it is critical to control for variations in territoriality, type of boat used, number of traps used, and size of investment. In this presentation on fishing skills, all of the information—including quantitative data—was obtained from 33 men in five adjacent communities in the central part of the coast. All of these communities are on open ocean, and all the territories are of the *nucleated* variety. All of the 33 men in our sample fish between 350 and 675 traps from inboard-powered boats ranging from twenty-eight feet to thirty-eight feet long. Their boats are equipped with hydraulic trap haulers and depth finders or recorders. None had radar. None of these 33 men employs a sternman for more than a few weeks a year, and most have none at all. While controls of this kind are not necessary to identify critical skills (Argument 2), they are absolutely essential to quantitatively link skills and individual output (Argument 3).

IV. ARGUMENT 2: CRITICAL FISHING SKILLS

The Views of Novices and Experienced Fishermen

Although fishermen repeatedly stated that skills were important, it was very difficult to ascertain exactly what these skills were. In the summer of 1973, I administered a questionnaire to a few men, but it quickly became apparent that formal interviewing techniques were not effective in studying skills. Questions designed to elicit data on skills were often only successful in eliciting instances of Maine humor. One of my questions was, "When you were just beginning to go lobstering, what was the most important thing you had to learn?" One man answered: "the rocks;" another humorist said "when to stay home;" still a third "to steer;" a fourth, somewhat bitingly, "where the highliners set their traps." While there is undoubtedly a serious element in all these answers, they really gave very few insights beyond the obvious.

One question got results worth reporting. I asked: "In the first five years you went lobstering, what single skill did you learn that helped you to increase your income most?" The answers are particularly illuminating if they are broken down according to the amount of experience these men have had in the fishery.[11] The results are recorded in Table 1.

TABLE 1 Critical Skills as Defined by Experienced and Inexperienced Lobster Fishermen

Number of Men		*Answer*
Over 5 yrs. Experience	Under 5 years Experience	
17	4	Where to place traps
2	3	How to build traps properly
1	2	How to cut losses and reduce costs of trap losses, boat damage, etc.
	1	"How to live with other men in the harbor gang. Avoid gear tangles."
	1	Proper engine and boat maintenance
1	1	Proper kind of bait
Total	33	

The men with under five years experience were concerned with such matters as how to build traps, maintenance, proper kind of bait, and so forth. Undoubtedly these are matters of concern to any fisherman, and are skills that anyone going in the business would have to learn. However, the experienced men overwhelmingly reported that knowing how to place traps was the major factor in their success. Closer questioning revealed that this was the skill that even experienced men found most difficult to master, and the one they thought made the most difference. It was also the skill that men were least willing to talk about, since, as one man put it, "You are asking the secret of how I earn my living." A good many months of research demonstrated that this was far from an exaggeration.

Of course, lobstermen must know how to pilot a boat, maintain equipment, build traps, and so forth. Undoubtedly, such matters are of primary importance to a beginner. But such skills are apparently relatively easily learned in comparison with the lifelong problem of learning where to place traps. Given the importance of trap placement to the fishermen, we will concentrate most of our attention on this issue.

Trap Placement

1. Novices and highliners. The overall pattern of trap movement is relatively simple. Men move their traps out into deeper water in the winter and back in shallower water in the warmer months of the year. Some of the younger men seem to know very little more than this. They memorize a set of "moves" and mechanically change the position of their traps without really understanding many of the factors involved. Certainly they understand none of the subtleties. Often, they merely follow an older man around and put their traps where he does—sometimes putting their traps literally on top of his. Naturally older men greatly resent this. At times they "cut off" the offending traps, but more often they resort to more subtle plays. Older men will often put traps in places they know are not productive merely to mislead novices. There are stories of experienced men tying buoys on pieces of concrete block to simulate a string of traps. On several occasions, I have heard of "highliners" putting gear dangerously close to shore when a storm was coming and moving it into deeper water at the last minute. Any novices lured into shallow water by these tactics are almost certain to lose a few traps. Still, the rewards of following a "highliner" are high enough that most novices are guilty of the practice at one time or another.

After awhile, most lobstermen learn to move their traps properly, but typically this takes several years. Some men never learn. Every harbor has at least one middle-aged lobsterman who just cannot seem to catch lobsters.

2. Reducing trap losses. Lobstermen try to exploit all the "lobster bottom" possible relative to trap losses, costs of fuel, costs of bait, and so on. In an attempt to limit trap losses, lobstermen take a good many factors into account. Traps are, of course, always placed in the traditional territory belonging to the fisherman's harbor gang save for those rare occasions when one is willing to risk traps for political gain. They are never placed in water so shallow that they would be exposed to air at low tide, nor are they placed in water over forty fathoms deep (240 feet) because the cost of forty fathoms of "warp" is prohibitive given the expected catch.

Fishermen also take into account several other factors which are not so obvious. One cannot put traps in water that is "too deep," or where the current is so swift that the trap buoy is pulled under water. Traps placed "over their head" are usually lost for good. For similar reasons, lobstermen must be careful not to place traps with relatively short warps near "deep holes" where the current might sweep them over the edge into deeper water. On the other hand, lobstermen cannot keep long warps on all their traps either. If the warp is too long, it will almost invariably become entangled with those of other men who have traps nearby. It can easily take a man five minutes and a great deal of aggravation to disentangle one trap from a few others.[12] Men try to keep the warps on their traps short enough to avoid entanglements. Exactly how much warp they leave depends on the current, number of traps nearby, depth, and so on.

In addition, trap placement depends on seasonal weather patterns. If traps are placed in shallow water too close to ledges, they will be destroyed if a storm comes along. Traps which may be perfectly safe in relatively calm weather can be destroyed by a storm's wave action pounding them against the bottom or sweeping them up on the beach. In general, men do not place their traps in shallow water or near ledges in the fall or the spring since these are the times when violent storms are apt to arise suddenly. But they do not completely avoid these areas either, for to do so would be to abandon some of the best lobstering areas available. The usual strategy is to place traps close enough to shallow water to catch lobsters, but in deep enough water to avoid dangerous wave action should a storm arise. Men sometimes place some of their traps in dangerous areas, planning to remove them if a storm approaches.

Some men are very successful at the game of "working around ledges," but novices typically lose a good many traps when they try it. I know of one young man who went into business in March 1973 and promptly put all his traps in shallow waters near unprotected ledges in April. While he caught a few more lobsters than the more cautious men, his trap losses were enormous. He was out of business within a year. There is also some statistical evidence that experience in "working around ledges" is a valuable asset. The men in our sample with over fifteen years experience lost only 9 percent of their traps in 1973, while the men with under five years experience lost about 17 percent of their traps from all "natural" causes.

Success in this game depends on keeping a "weather eye peeled" and on the willingness to work long hours moving traps when storms are pending. The trap carrying-capacity of the boat is also a factor taken into account, since men with bigger boats can obviously rescue many more traps when storms are impending than men with smaller boats.

Men also have to learn to reduce trap losses (relative to the probability of increasing total catch) by being very careful about placing traps in ship channels or in places where a lot of dragging is going on. In these places, warps can be cut by propellors, and traps destroyed by nets.

The importance of political factors in trap placement cannot be underestimated. One has to learn to keep a distance from some men. Exactly how close one can crowd a neighbor depends on political factors such as a man's own standing in the harbor gang and his relationship with the other man. In addition, the men of a harbor gang always keep traps on the periphery of their harbor area to maintain their claim over it. In fact, many times they will deliberately place traps over their line by a few yards in an attempt to enlarge their area. This is called "pushing the lines." Moreover, even within a given territory of a harbor gang men will place traps in areas where they do not expect to catch lobsters to reserve those areas for a time when lobster can be caught there. This process of "camping out," as it is known, is especially prevalent in the Casco Bay area where escalation in the number of traps used has produced severe gear congestion.

The most important skill necessary to reduce trap losses is a knowledge of the depth of water in the entire area of the harbor gang. In the past, this was the most difficult skill to acquire since one could only learn the depth by using a hand line or by noticing how much warp was necessary in various spots. The task was especially difficult far from shore, where one could only mark locations by reference to objects on shore. The task of memorizing water depths was made easier by the fact that lobstermen had far smaller territories then; but, even so, it was a formidable one. Some men resorted to making small maps, but most tried to recall the pattern of the bottom. Thus, they would talk about "trenches" or "rock areas." At present, men still memorize the depths in certain areas, but they also make extensive use of charts, depth finders, and depth recorders. In fact, a depth finder is regarded as basic equipment by commercial fishermen; and the use of this equipment has made it much easier to learn the "bottom" and has greatly reduced mistakes in trap placement. Nevertheless, merely learning the depths is still no easy task. While a young fisherman has an easy time finding out what the depth is in any particular location, he still may not have the experience and judgement to use this information to reduce his trap losses.

3. Trap placement and the biology of the lobster. Placing traps where they will produce is far more complicated than might first be imagined, for lobsters are not randomly scattered over the bottom. Furthermore, the productivity of a given small area is very much dependent on the amount of fishing effort previously expended there and the time of year.

Even though fishermen and fisheries biologists do not agree on many issues (e.g. stock assessment), they are in agreement on many of the factors concerning the micro-ecology of the lobster, particularly as these factors affect catch. With rare exception, there is little which "skilled" fishermen can learn from fisheries' biologists to increase their output. (The exception may be issues concerning migration and water temperature.)

The lobster is a nocturnal animal that lives in water between one and 200 fathoms. The vast majority live in rocky areas where they find protection from predators by hiding in crevices. Even though lobsters have the capacity to burrow in mud and sand, smaller numbers are found on muddy or sandy bottom. In this regard, J. S. Cobb reports, "Here 52 lobsters were seen in a 530–m^2 survey area on rocky bottom, giving an average of one lobster every 10.2–m^2. No lobster were seen in a 530–m^2 area on sandy bottom, and one lobster burrowed under a small rock outcrop was seen in 530–m^2 of mud bottom surveyed" (Cobb 1971:109). Since most lobstermen are fully aware of these facts, most try to place their traps on rocky bottom. When a fisherman talks about "good lobster bottom" he is referring to areas where there are a lot of rocks—preferably piles of rocks—or places where the rocks meet muddy bottom. Up until 1959–60, the smaller number of lobsters living on muddy or sandy bottom was left relatively unexploited since no one deliberately put traps in such areas. Besides being relatively unproductive, muddy areas are usually far deeper than the rocky mounds, and require traps with far more warp to exploit them. Today, increasing numbers of good fishermen are setting larger numbers of traps "on the mud." Two factors appear responsible for this change. First, given the general escalation in number of traps used, there is no longer enough room to place all traps on "good bottom." Second, the hydraulic hauler has decreased the amount of energy and time necessary to exploit deeper, muddy areas.

In the past, it was relatively difficult to memorize all the types of bottom, for exactly the same reasons it was difficult to learn the depths. One could only tell if the bottom was muddy or rocky by the feel of a lead weight as it was bounced off the bottom, or by noticing if there was mud on the bottom of traps when they were pulled. Of course, this knowledge about a location would do a man little good in the future unless he could find the exact spot again. The use of depth finders and recorders allows lobstermen to identify the type of bottom with reasonable certainty. Muddy bottom gives off a much weaker echo than rocky bottom, thus showing up on the depth recorder paper in the relative darkness of the markings, and on the depth finder in the strength of the light signals. The signals indicating rocks or mud are quite noticeable to the experienced eye. While these facts are known to everyone in the fishery, skilled men still memorize good lobstering areas, and do not depend on the machine completely. Not only can they find the larger, well known, areas faster than unskilled men, they also know about and can relocate relatively small features—a particular "mud hole" or a small rocky mount. Some men have become so skilled with the depth equipment, that they can locate a wreck of a boat or even auto bodies.[13]

In addition, some men have a superior knowledge about certain aspects of the biology of the lobster. Some men, for example, are aware that there are no lobster in shallow areas where one can see "white, bald rocks." The few lobstermen I have talked to about this subject seem to think that this is due to the fact that lobsters do not inhabit areas where there is no kelp to hide in. There is another possibility. Mann and Breen (1972:603–04) have demonstrated that in areas where the lobster is overexploited by man, the population of sea urchins expands rapidly, since there are not enough lobsters to control their population. The burgeoning population of sea urchins, in turn, overgrazes the kelp. In essence, if this view is correct, the bald rocks may indicate nothing but overexploitation by man.

Furthermore, some skilled lobstermen are aware that lobsters tend to live in places where the bottom is irregular. Rocky areas with a lot of mounds and rock piles are far more productive of lobsters than flat rocky areas. Muddy bottom with occasional rocks and breaks in the contour lines are more productive than flat muddy or sandy surfaces.

> My survey at Bonnet Point where rocky, sandy, and muddy bottoms existed within 100m of each other showed that segregation according to substrate occurs over short distances as well. A suitable substrate need not be rocky but simply one where shelters are available or can be constructed as indicated by lobsters inhabiting the muddy bank at Castle Cove. The environments inhabited by lobsters show a discontinuity in bottom relief, and my impression is that, in general, the more gross the discontinuity in bottom relief, the larger the lobsters (Cobb 1971:113).

When experienced lobstermen are fishing water over 5 fathoms deep where they cannot see the bottom, they use their depth recording equipment to locate small hummocks or places where the bottom banks off rapidly. Placing traps on such locations takes a good deal of practice. An experienced lobsterman usually has a very good idea where such discontinuities in the bottom are. Sometimes he will circle around in a 100 yard area intently looking at his depth recorder. When the recorder line wavers slightly, the fisherman will rapidly turn the wheel and run back over the spot to dump his trap, taking into account the effect wind, tide, waves, and current will have on its ultimate landing place. Every time the trap is pulled, he will reposition it according to the catch of lobsters he is obtaining from it. When a lobsterman can see the bottom, he will attempt to place traps near very large boulders or in crevices in the rocks.

It needs to be stressed that skilled lobstermen place traps with pinpoint accuracy. One lobsterman I know very well deliberately keeps a trap in a large crevice within ten feet of shore. He says he gets at least one large lobster out of that trap every day between mid-July and mid-August. I saw him take two pound lobsters out of that trap three days in a row. It took me some time to fully appreciate the incredible piloting skill necessary to maneuver around the offshore ledges, white with breaking surf, to get that large bouncing boat next to that crevice. Even if a novice fisherman knew about this crevice and could find it, he would never dare take his boat into this shallow area. One miscalculation of wind or tide and he could hit a ledge with sufficient force to severely damage the hull of his boat.

Both fishermen and fisheries biologists agree that lobsters are much easier to catch in the winter in deep water, and in summer in shallow water. Their explanations for such seasonal variations differ radically, however. The biologists believe that the major factor influencing where lobsters can be caught is water temperature. The fishermen see this as a matter of migration.

The fishermen generally agree that the lobsters migrate into deep water in the cold months of the year, and then slowly migrate back towards the shore in the spring. In the early summer, they say, lobsters come within 100 yards of shore to molt in the rocks. They move their traps accordingly. "The trick," as several highliners have phrased it, "is to just keep ahead of them." That is, a man tries to place his traps in places where lobsters are going to be in the next few days, rather than behind the lobster migration or in areas which lobsters may migrate out of in the next day or two. By placing traps in the path of the "migration" a lobsterman is accomplishing two things. First, if a man can't pull his traps for a few days due to bad weather, the traps will still be productive. Second, by keeping his "strings" ahead of the lobsters (and the other fishermen) he is able to reserve the choice trap locations for himself.

According to several fishermen, lobsters do not move gradually over the bottom. Sometimes one may be able to keep a string of traps in virtually the same location for two weeks or longer; at other times one has to move them almost constantly to "keep up" with the lobsters. The best fishermen try to predict where lobsters can be caught by using test traps. They will put a few traps in areas lobsters have not "migrated into" and will keep others behind. When the "rear" traps no longer catch anything but undersize lobsters, and the "lead" traps begin to produce, they will move their string closer to the "lead traps." Sometimes strings of traps are laid out in long rows parallel to shore, but often skilled lobstermen will cluster traps in seeming helter skelter confusion in a fairly wide band of water. This is due to the fact that they are identifying "productive lobster bottom," "lead traps" and "rear traps" more in terms of depth than in terms of physical distance from shore. In fact, this trait seems to distinguish "skilled" fishermen from novices. The skilled men are trying to place traps on "good lobster bottom" (i.e. rocky bottom, etc.) at particular depths, regardless of the geometry of trap placement. The novice tends to think in terms of placing traps a certain distance from shore in long strings where he can keep track of them better. Considerations of precise depths and the type of bottom are of secondary importance to him.

While fisheries biologists admit that traps must be moved seasonally to maximize the catch, they strongly disagree with the idea that lobsters have patterned, seasonal migrations. According to biologists, the results of several tagging experiments show conclusively that lobsters live throughout their entire lives in one small area and do not move seasonally (Harriman 1952:2; Wilder 1954;9; Wilder and Murray 1956, 1958).

> As far as seasonal migration was concerned none could be determined from the data. There was comparatively little movement of any kind. Among the 570 lobsters tagged in 1950, only 18 lobsters moved an estimated one-quarter mile or more from the place of planting. Of these, six moved at least two miles (Harriman 1952:2).

The "day to day and seasonal differences in the catch" are explained by biologists in terms of temperature differences (Wilder 1954:9). There is, they assert, a relationship between water temperature, general activity of the lobster, and the rate at which they are caught (Paloheimo 1963:62ff). Thus, from the point of view of the biologist, lobsters are most easily caught in the winter months in deep water where there are relatively warm currents; in the summer, in the warm water near shore.[14]

Despite this body of scientific evidence, there is some recent data which tend to support the arguments of the fishermen. Recent tagging experiments on the outer continental shelf indicate the lobsters in those areas do undertake extensive seasonal migrations. "The distribution of the recoveries demonstrated shoalward migration in spring and summer and a return to the edge of the shelf in fall and winter" (Cooper and Uzman 1971:288). The distances travelled were surprisingly large.

> Twenty-one percent of the recaptured lobsters had moved distances less than 16 km, 58 percent between 16 and 80 km, and 21 percent in excess of 80 km./ . . . The apparent rate of travel ranges up to 10 km/day (Cooper and Uzmann 1971:291).

While no migrations have been recorded by biologists inshore, these data suggest that the fishermen's claim that lobsters migrate may be scientifically verifiable.

One might argue that it makes little difference whether the fishermen or the biologists are correct. Both would certainly agree on the way traps have to be moved seasonally. In fact, it might be demonstrated that the fishermen are following a temperature gradient, even though they are doing so for the wrong reasons.

In summary, the skills lobstermen identified as most important are associated with trap placement. In placing traps, lobstermen take a wide variety of variables into account which are designed to maximize catch relative to cost of gasoline, bait, and—most important—trap losses. Exactly where a man places traps depends on factors such as depth, season of the year, probability of storms and their direction, size of boat, local trawling and dragging activity, and several types of political factors, as well as lobster migration activity, type of bottom, state of the kelp, and so forth. I did not fully appreciate the complexity of the problem of trap placement until I spent a day on a boat with an experienced fisherman who was in the habit of talking to himself. I was amazed at the number of factors he was simultaneously taking into account.

V. ARGUMENT 3: FISHING SKILLS AND INDIVIDUAL OUTPUT

Most fishermen are fully aware of all the various factors involved in trap placement, and know that novices and skilled men place traps in very different ways. Many have developed some kind of shorthand way of describing these differences. For example, one old man is fond of saying "the old-timers know how to make their traps count; the kids are just throwing them all over the bay." Several other men talk about these differences by using the analogy of saturation bombing versus pinpoint bombing.

In order to test out the fishermen's hypotheses concerning the importance of "pinpoint" trap placement, I gathered a great deal of data on 33 men, including data on catches, number of traps, and size of lobsters caught. (In section II, it was pointed out that these men were chosen with a view towards controlling for several different kinds of economic, ecological, technological, and institutional factors.) An analysis of these data indicates that "pinpoint" trap placement pays very well. These data are summarized in Table 2.

TABLE 2 Output of Lobsters by Men of Different Skill Levels in August 1973

	Pinpoint trap placement	*Unpatterned trap placement strategies*	*Saturation trap placement*
No. of Cases[15]	8	14	11
Catch	220 lbs/day	165 lbs/day	140 lbs/day
Mean carapace length of lobsters caught[16]	90.2 mm (2674)	89.0 mm (n 820)	88.9 (n 1385)
Mean no. of traps in water	365	490	330
Gross income from lobstering 1973[17]	$18,640.00	$15,684.00	$14,555.00
Mean age of fishermen	46.1	44.1	33.7
Mean traps pulled per day	162	165	150
Pounds caught per trap	.97	.83	.65

As one might expect, the men who used "pinpoint trapping" techniques were older than the men who were not placing their traps carefully. However, two of the men who "pinpointed" their traps were under 30, while four of the men who merely saturated the water with traps were over 35. This again indicates that skill is not merely a matter of age alone.

It is important to notice that the "skilled" men were not expending much more effort than the "unskilled" men. They had only a few more traps in the water, and were not pulling that many more traps per day. Nevertheless, the "skilled" men got far greater returns for their efforts than the "unskilled" men. Their gross income from lobstering was an average of $4,085.00 higher. Despite the low number of cases, the difference in mean income is statistically significant.[18]

Moreover, there is a great difference in the mean pounds of lobster caught per day. Usually one cannot assess a man's skill in placing traps by merely noting daily catch because such a measure does not take into account the number of traps fished, or the length of time the traps are in the water. Two men who catch the same number of pounds of lobster are not equally skillful if one is fishing twice as many traps as the other. Ordinarily, the number of pounds of lobster caught per trap is a much better indicator of skills. In this case, however, the difference in pounds caught per day probably does indicate a great difference in skills, since the fishing effort of these two groups of men is so similar.

There is also a great difference in the pounds of lobster caught per trap by these two different sets of men. The men who carefully pinpoint the location of their traps caught an average of .97 pounds per trap in August, 1973; while the 11 men who scattered their traps caught only .65 pounds per trap.

Last, there is a difference in the size of lobsters caught by these three sets of men. The fact that the "skilled men" get slightly larger lobsters is probably attributable to the fact that they place many more traps near large rocks and crevices where larger lobsters can be caught. In any case, this difference means that the "skilled" men are catching a higher proportion of large "dinner" lobsters bringing a higher price per pound. This undoubtedly is reflected in the difference in incomes.

VI. OTHER SKILLS AND STRATEGY

There are several other factors linked to technical skills that probably have some influence on output in the lobster industry. First, many men claim that the kind of bait used strongly influences catch. Unfortunately, there appear to be more myths and rumors about bait than about almost any other aspect of the industry. Fishermen comment on freshness or rancidness of bait, the amount of salt used, seasonal preferences of lobsters, amount of oil in the fish, and so on. Three "highliners" in the sample claim that a combination of redfish frames and bagged herring will produce the best results in the late summer and early fall.[19] They may well be correct. Unfortunately, I do not have enough data on type of bait used to test this hypothesis with any hope of statistical reliability.

Second, knowing when to pull traps can apparently have a great effect on catch. If a trap is pulled a few hours after it has been set, the results are poor, because the bait has not had a chance to "work." However, one cannot wait too long either, for there is a maximum number of lobsters that any trap will catch without being rebaited. (There is strong evidence that the marginal productivity of a trap decreases after a few days [Thomas 1973:41].) Exactly when a man pulls his traps depends on the weather, the number of traps used, and the time of year. In the winter, when lobsters are moving more slowly and the bait lasts longer, one can leave a trap in the water as long as seven or eight days with favorable results. In the summer, a trap catches all it is going to in three or four days; most men plan to pull them more often than that.

Third, many men argue that traps constructed with certain features will catch more lobsters. While the number of theories concerning the way to build a "proper lobster trap" are legion, there is some evidence that "venting a trap" does have favorable results. Basically "venting" refers to the practice of leaving a large enough space between the bottom lathes on a trap to allow undersize lobsters to escape. If traps are "unvented," a small lobster in the trap will defend the trap when a larger lobster approaches. If the trap is "vented" the small lobsters can escape, leaving the trap free for legal animals to enter. A study of catches would almost certainly reveal that "vented" traps catch more legal lobsters than the "unvented" ones.

Last, given any level of skills, there may be differences in the way different men estimate risks. This is probably tied up with the type of equipment they have, their "goals," and the marginal utility of income for them.

While a great deal more information is needed to assess the effect of risk on output, it should be noted that men of approximately equal experience are not employing the same strategies. Many "highliners" move their traps in the same general pattern as the vast majority of the men—inward in warm weather, and into deeper water in the colder months. But some highly experienced men are not operating this way. One older man, for example, keeps many of his traps in deep water throughout the year. Even in the middle of the summer, he is pulling traps in 30 fathoms of water. He says "there are not as many lobsters out deep at this time of year," and that the waves "are bigger," but his gear is not always being tangled up with those of other men, and he has virtually no competition for the "remaining" lobsters in this area. In addition, he loses very few traps in the bad weather. This man is placing a very high value on conflict avoidance and on minimizing trap losses. He may be willing to trade off some income to attain these goals. However, given the fact that he is fishing these waters virtually alone, this strategy may be one that maximizes catch. During August, 1973, this man caught an average of 1.1 pounds of lobster per trap pulled (slightly higher than the average "skilled" fisherman). Another older man fishes only 150 traps with a very small twenty-six foot boat. He deliberately places all his traps in "shoal water" (under ten fathoms) throughout the year. Although his trap losses are very high, he too has very little competition much of the time. His strategy, apparently, is to avoid competition and to take advantage of the maneuvering ability of his boat. He is willing to sacrifice traps to attain these ends.

The overall strategy of any lobsterman is to maximize catch relative to costs of labor, bait, gas, and trap losses. There are, however, a number of different ways one can operate to achieve this end at any given level of skill.

VII. SKILL AND TECHNICAL CHANGE

Technical change in the lobster industry would likely produce a levelling effect, and could reduce much of the advantage "skilled" men now enjoy. In any industry, the type of skills required is linked to the particular technology in use and to a view of the world concerning the reasons that technology is effective. Given the technology currently used in the Maine lobster industry, skills in trap placement are critically important. The way traps are placed is connected to the idea that lobsters migrate seasonally, are found in rocky areas, and so on. In this situation, older men who have had time to memorize depth, types of bottom, and lobster migration patterns, usually have a clear advantage over men with fewer years experience, particularly in the absence of institutional mechanisms to transmit skills. To be sure, the advent of the depth sounder has made it much easier to learn depths and types of bottom, but the instrumentation is inexact enough that memorization of locations of physical features is still important.

It is very unlikely that technological change would aid the men who are now considered "skilled" and help them maintain their advantage over the men who are now considered "unskilled" novices. First, if a new technology came into use, a premium would be placed on some very different types of skills. For example, if fishermen became convinced that lobsters near shore really are essentially nonmigratory, and are strongly affected by changes in temperature, it might be profitable for them to pay attention to changes in water temperature. Already, at least two lobstermen are beginning to experiment with temperature recording equipment, and are talking about buying more sophisticated electronic equipment. If these experiments prove successful, they may foretell a change to a whole new technology demanding skills in electronic maintenance, operation of temperature recording devices, knowledge of currents and thermoclines, and so on. The men who put so much time and effort into memorizing features of the bottom might find their skills antiquated, if not useless. Where this new technology is concerned, they might not be far ahead of novices.

Second, in industrial nations, technical change emanates from the scientific community to industry. There is little reason to assume that it would happen otherwise here. Of course, knowledge from the scientific community would be equally available to everyone in the lobster industry by direct or indirect means. Technological changes involving sophisticated equipment might even give the advantage to men with better education who are most used to learning new ideas via the written word. Thus, in the short run, technical change would tend to equalize the skills factor and bring fishermen back to a common starting line. In the long run, it would probably produce a redefinition of critical "skills" and a different set of skilled and unskilled fishermen.

NOTES

[1]Lee, for example, notes that a high proportion of all the animals taken by the Kung Bushmen, are killed by a small proportion of the population (Lee 1968:36–37). Accounts of northern trapping mention the fact that different trappers catch very different numbers of fur-bearing animals, and earn vastly different amounts from their trapping efforts (e.g. Nelson 1973:167; Savishinsky 1970:478–479; 483). A number of anthropologists have linked success in hunting to "skills" (Henriksen 1973:39; Service 1966:51). Recently, Richard Nelson has attempted to pinpoint some of these technical skills among both the Eskimo and Kutchin (1969, 1973).

[2]Recently, at least, two researchers have come to suspect that technical skills are of critical importance in several New England fisheries. Mueller has attempted to do a statistical analysis of the factors influencing expected catch in the Northern Shrimp industry. He is convinced that the results of his multiple regression analysis of the production function would be far better if he could get a surrogate variable for technical skill (Mueller 1975). In addition, Peterson has recently discovered that the gross income of New Bedford (Massachusetts) trawlermen varies greatly with ethnicity. She suspects that the success of the Norwegian captains is due to the fact that they have certain skills which they are passing on only to other members of their ethnic group (Peterson 1975).

[3]They also often say that they failed because of poor or inadequate equipment, and because they were driven out of business by the established men.

[4]The results of the first survey have been reported by Huq (1972).

[5]Trap days are computed by multiplying the number of traps a man has in the water by the number of days they are used.

[6]A standard two-tailed test of significance demonstrated that these results were significant at the .01 level.

[7]The federal government has made no effort to regulate the lobster industry, and the state of Maine has passed only a few conservation laws. To be legally taken, a lobster must have a carapace length of over 3–3/16 inches and under 5 inches. Regardless of size, breeding females are protected by a law requiring any lobsterman who catches an egg-bearing female to cut two notches in her tail flipper and throw her back. "Notched tail" lobsters may never be taken. In addition, the state has a licensing requirement, and requires lobstermen to mark their traps and buoys with tags and distinctive colors.

[8]However, in those same years the catch dropped from approximately 25 million pounds to 17 million pounds (Dow, Bell and Harriman 1973:45).

[9]Throughout the 1960s, there were approximately 6000 lobstermen in Maine. In 1973 about 10,400 lobster licenses were issued in the state. About 2800 of these men were full-time fishermen, the remainder being "part-timers" who earn most of their income ashore.

[10]Fishermen themselves have no universal name for the groups of men fishing from one harbor. They refer to the "Friendship gang" or the "Boothbay boys," and so on. I refer to these groups as "harbor gangs," although this term is rarely used by the fishermen themselves.

[11]The wording of these answers is essentially mine. No two men answered the question in quite the same way. Often this question would elicit a half hour monologue.

[12]After a storm it may take one or two days of steady work by everyone in the harbor gang to disentangle literally mountains of gear.

[13]Auto bodies are sometimes sunk on purpose to discourage trawling and dragging in certain locations.

[14]In all areas, lobsters are less active in the winter than the summer.

[15]Of the 33 men in our sample, 8 men clearly placed their traps carefully (pinpoint trap placement); 11 clearly did not (saturation trap placement). The other 14 did not clearly fall into either category. The 33 men were placed in one of these three categories on the basis of observation of their fishing techniques, and other men's assessment of them.

[16]In marine biology, lobsters are traditionally measured on the carapace—from the eye socket to the back of the body.

[17]Accurate information on income could only be obtained from 18 of the 33 men.

[18]A t-test was run to discover whether the differences in means was significant. The value of the t was 5.1 (p less than .001).

[19]Local processors fillet the redfish and sell the frames with heads and tails for lobster bait. Herring heads and tails are obtained from canneries, and are packed into small nylon net "bags" by the fishermen. To bait a trap, the lobsterman suspends the frames or the bags in the middle of the trap behind the "head."

LITERATURE CITED

Acheson, James M.

1975a The Lobster Fiefs: Economic and Ecological Effects of Territoriality in the Maine Lobster Industry. Human Ecology 3(3):183–207.

1975b Fisheries Management and Cultural Context: The Case of the Maine Lobster Fishery. Transactions of the American Fisheries Society 104(4):653–668.

Becker, Gary

1964 Human Capital: A Theoretical and Empirical Analysis with Special Reference to Education. New York: Columbia University Press.

Cobb, J. Stanley

1971 The Shelter-Related Behavior of the Lobster, *Homarus Americanus.* Ecology 52(1):108–114.

Cooper, Richard A. and Joseph R. Uzmann

1971 Migration and Growth of Deep-Sea Lobsters, *Homarus Americanus.* Science 171:288–290.

Davenport, William H.

1960 Jamaican Fishing: A Game Theory Analysis. Yale University Publications in Anthropology 59:3–11. New Haven, Conn.: Yale University Publications.

Dow, Robert L.

1961 Some Factors Influencing Maine Lobster Landings. Commercial Fisheries Review 23(9):1–11.

Dow, Robert L., Frederick W. Bell and Donald M. Harriman

1973 Bio-Economic Relationships for the Maine American Lobster Fishery with Consideration of Alternative Management Schemes. U.S. National Marine Fisheries Service, File Manuscript No. 149.

Harriman, Donald M.

1952 Progress Report on Monhegan Tagging—1951–52. Maine Department of Sea and Shore Fisheries. Mimeo.

Harris, Marvin

1968 The Rise of Anthropological Theory. New York: Thomas J. Crowell Company.

Henriksen, Georg

1973 Hunters in the Barrens: The Naskapi on the Edge of the White Man's World. Institute of Social and Economic Research, Newfoundland Social and Economic Studies No. 12. St. John, Newfoundland: Memorial University of Newfoundland.

Huq, A. M.
1972 A Study of the Socio-Economic Impact of Changes in the Harvesting Labor Force in the Maine Lobster Industry. U.S. National Marine Fisheries Service, Economic Research Laboratory, File Manuscript No. 110.

Lee, Richard B.
1968 What Hunters Do For a Living, or, How to Make Out on Scarce Resources. *In* Man the Hunter. R. B. Lee and Irven DeVore, eds. Pp. 30–48. Chicago: Aldine.

Mann, K. H. and P. A. Breen
1972 The Relation Between Lobster Abundance, Sea Urchins and Kelp Beds. Journal of the Fisheries Research Board of Canada 29:603–609.

Mueller, Joseph
1975 Personal Communication.

Nelson, Richard K.
1973 Hunters of the Northern Forest. Chicago: University of Chicago Press.
1969 Hunters of the Northern Ice. Chicago: University of Chicago Press.

Paloheimo, J. E.
1963 Estimation of Catchabilities and Population Sizes of Lobsters. Journal of the Fisheries Research Board of Canada 20(1):59–88.

Peterson, Susan
1975 Personal Communication.

Proskie, John
1971 Costs and Earnings of Selected Fishing Enterprises: Atlantic Provinces 1968. Primary Industry Studies 1(18). Ottawa: Economics Branch, Fisheries Service Department of Fisheries and Forestry.

Savishinsky, Joel S.
1970 Stress and Mobility in an Arctic Community: The Hare Indians of Colville Lake, Northwest Territories. Unpublished Ph.D. dissertation, Cornell University.

Scherer, F. M.
1970 Industrial Market Structure and Economic Performance. Chicago: Rand McNally Co.

Service, Elman R.
1966 The Hunters. Englewood Cliffs, N.J.: Prentice-Hall.

Thomas, James C.
1973 An Analysis of the Commercial Lobster (*Homarus Americanus*) Fishery Along the Coast of Maine, August 1966 Through December 1970. U.S. Department of Commerce, National Oceanic and Atmospheric Administration, National Marine Fisheries Service, NOAA Technical Report NMFS SSRF–667.

Wadel, Cato
1972 Capitalization and Ownership: The Persistence of Fisherman-Ownership in the Norwegian Fishery. *In* North Atlantic Fishermen. Raoul Anderson & Cato Wadel, eds. Institute of Social and Economic Research, Newfoundland Social and Economic Study No. 5, St. John, Newfoundland: Memorial University of Newfoundland.

Wilder, D. G.
1954 The Lobster Fishery of the Southern Gulf of St. Lawrence. Canada, Fisheries Review Board of Canada, Atlantic Biological Station, St. Andrews, N. B. General Circular No. 24.

Wilder, D. G. and R. C. Murray
1958 Do Lobsters Move Offshore and Onshore in the Fall and Spring? Fisheries Research Board of Canada, Atlantic Progress Report No. 69:12–15.
1956 Movements and Growth of Lobsters in Egmont Bay, P. E. I. Fisheries Research Board of Canada, Atlantic Progress Report No. 64:3–9.

6

The Technology of Irrigation in a New Mexico Pueblo

RICHARD I. FORD

Museum of Anthropology, University of Michigan, Ann Arbor, Michigan

THE IRRIGATION HYPOTHESIS

Social scientists and political historians have been fascinated by the social consequences of the cultural manipulation of water resources for practical ends. All agree that the artificial delivery of water to cultivated plants is an important circumvention of Nature's limitations, and that the more complicated irrigation becomes, the more complex is the related social system. But these generalizations give way to scientific debate over causality.

Wittfogel (1957), in particular, has envisaged the origin of a complex social type, the hydraulic state, as resulting from the institutionalization of administrative authority to regulate large scale water-works. The assumed priority of hydraulic agriculture over the institution of the state has been challenged by archaeologists (cf. Wright 1976). Today this technology is interpreted at least by most anthropologists as an enabling mechanism for maintaining the state (Service 1975:273–275).

Even though the postulated importance of irrigation in the rise of civilization has been lessened, it is still conceived of as an important form of technology that shapes entire social systems. Again, the intellectual influence of Wittfogel is evident. He and Goldfrank, based on Goldfrank's field observations in the Pueblo southwest, concluded that the Rio Grande Pueblos evolved their centralized village authority because of the discipline required to maintain their irrigation system (Wittfogel and Goldfrank 1943). Following this suggestion Eggan (1966:130, 138) considered canal irrigation as a major variable accounting for the difference in community organization between the clan-lineage based Western Pueblos and the dual organization-sodality organized Rio Grande Pueblos. This theme, dubbed the irrigation hypothesis by Dozier and further developed in a series of seminal books and papers (Dozier 1960:150–51; 1970:131–33), remains the accepted explanation for Eastern Pueblo social organization.

The irrigation hypothesis states that the presence of hydraulic agriculture in the Pueblos of the Rio Grande Valley and its tributaries necessitates highly coordinated community activity directed by forceful leaders who can, through religious sanction and social authority, mobilize manpower to maintain the canals and allocate accessibility to the precious water. In contrast, the clans of the Western Pueblos are interpreted as inadequate for the coordination of a large work force to satisfy the technological demands of irrigation. If irrigation does account for this geographically based structural dichotomy, then evidence should show that water control in the Western Pueblo area is technically less complex than in the East.

Recent archaeological research demonstrates that this distinction is incorrect. Canal irrigation, although pre-Hispanic in origin in the Rio Grande Valley, is not limited to that river drainage alone. Furthermore, the canals, reservoirs, and terraces in the prehistoric Western Pueblo region are far more complicated than any which have been unearthed in the Rio Grande area (Vivian 1974). While these discoveries vitiate the distinction between the two areas, they have not been associated successfully with any particular prehistoric social organization. Therefore only a thorough study of the technology of irrigation will validate or dispute the basic assumptions of the hydraulic hypothesis.

The study of technology is an examination of the processes required to transform the biological and physical environment into useful things. From this perspective each technological activity is goal-oriented; its function is to satisfy human needs by altering the material world. The form of the final product is the outcome of more than simply a combination of tools and ability. It is a result of cultural traditions and social relations for organizing work. Skills, knowledge, and procedures are taught and learned in a social context and conducted in accordance with beliefs about the world. Culture provides the social rewards and incentives, the production organization, and the ideological constraints that shape the finished product, although the properties of raw materials and the physical limits of technological processes are equally important. Purposeful physical change is the basis of technology. A dibble is a modified limb; a pot is transformed clay; an irrigation canal is an alteration of the natural landscape. Each can be studied by means of the laws of physics, chemistry, and biology. This important recognition permits the construction of models based upon inherent properties of the materials and then comparison of these with the reports of field research describing the actions of the craftsmen and workers themselves.

To understand irrigation technologically we must examine the capacity of the ditch, rates of flow, availability of water, methods of spreading water, the needs of crops, sources of water loss, and so on. Each is a measurable variable that permits the ascertainment of the limits of the physical procedures and materials and thus a determination of how a culture copes with them in satisfying its requirements. For example, a canal carries only a certain amount of water each day but the amount used will vary with the number of farmers, the number of hours they work, and the condition of the soil and plants.

Since the irrigation hypothesis purports to apply to all Eastern Pueblos, a detailed study of one, following the procedure outlined briefly above, should provide a test of the idea and suggest how technological studies can inform social understanding. Picuris Pueblo, a northern Tiwa speaking population, was selected for an investigation of irrigation technology.

IRRIGATION IN A PUEBLO SOCIETY

Picuris Pueblo

Built on a 700 year archaeological foundation, this adobe village, once home to perhaps a thousand residents,[1] is located between Santa Fe and Taos, New Mexico, along a tributary of the Rio Grande. Disease, conflict, and migration reduced this population drastically, and the descendents have numbered about 100 throughout the American presence in the Southwest. The haunting spectre of history brought more contempt than mercy from Indian agents as they annually maligned these often isolated and victimized Indians. Anthropologists have not enhanced their reputation either. Until the recent work of Dick (unpublished) and Brown (1973) anthropologists classified Picuris as "Mexicanized" or at best as a notorious example of social disorganization. The community deserves better, and its persistent survival in a hostile environment can teach anthropologists a great deal.

Picuris is situated at an elevation of 7,300 feet on a Pleistocene pediment flanking the southern slope of the Picuris Mountains, a spur of the Sangre de Cristos, which bound the north side of the pueblo at heights of over 10,000 feet. The Rio Pueblo carries run-off and eroded sediment from the same mountains westward through the Picuris Valley and eventually to the Rio Grande at Embudo. This river is the source of irrigation water and its narrow flood plain provides most of the arable land.

Under normal climatic conditions a restricted number of cultigens can be grown, but the Mountain People must adapt to unexpected variations, not average conditions.[2] The mean annual temperature is 47° F and the growing season (frost-free) lasts 128 days. However, these figures are deceptive since the last killing frost in the spring may occur in June and the first severe frost can strike in late August. The close proximity of the Picuris Mountains makes cold air drainage an additional problem. Winter temperatures have dipped down to -40° F. The 15.19 inches of annual precipitation is likewise misleading. In actuality the monthly pattern of effective precipitation (moisture available for plant utilization) is very critical. In May 1.3 inches are available, while in June only 0.94 inches are useful. Normally the amount increases significantly in July and in August. These figures contrast with the consumptive-use by the plants for the same months: May, 2.69 inches; June, 4.40 inches; July, 5.46 inches; August, 5.06 inches; and September, 3.52 inches.[3] In other words, because of insufficient rainfall in this semi-arid land, irrigation is essential, and the depth of the melting snow pack in the mountains becomes critical. The availability of this source of water is rarely a problem, however, as Frederick Müller (1893:118) discovered when he encountered 4.5 feet of snow while crossing the mountains from Taos on February 26, 1891.

The demographic characteristics and culture described here apply to the closing decade of the last century when all able-bodied men were full-time farmers. In 1890 Picuris housed 108 inhabitants. Although males outnumbered females 62 to 46, the total population over 18 was 61 while there were 30 heads of family, suggesting that the skewed sex ratio occurred in the under 18 cohorts (Donaldson 1893:92).

The household, variously extended, is the basic domestic work and consumption social unit in Picuris. The adobe house and various outbuildings and corrals are usually owned by the adult male but women may inherit houses as well. Other property, including land, is received equally by men and women from parents and other relatives. Households gather with their bilateral relatives on various occasions, especially to commemorate rites of passage.

Pueblo residents belong to pan-village sodalities. Everyone joins his father's North Side or South Side ceremonial groups. The other sodalities have been classified by Brown (1973). Four societies are named after the seasons and through ritual they regulate all life during the season of their namesake. In addition, there is a Water Clown society. Special powers relating to man and animals are vested in a Scalp Headman and a Hunt Headman. Picuris has a village Headman, or cacique, who together with the nine heads of the other societies forms the Council of Principles (Brown 1973: 100–106).

Until recently when all adult males became the government, the Council of Principles conducted the ceremonial affairs of Picuris and appointed the Spanish imposed secular officers. Important among the latter group are the annually appointed governor and the *mayordomo,*[4] and his assistant. The Council, governor, and *mayordomo,* are directly concerned with the management of the irrigation system.

Description and Importance of Irrigation

The age of the Picuris ditch network is unknown. Centuries of excavation and repair mask the critical evidence needed to verify the folk belief of its prehistoric origin. Although in some exceptional years aboriginal crops could have been grown without irrigation water, the use of this technology reduces plant stress and increases yield.

The present main ditch has been modified during this century to a length of at least a half mile longer than the approximately 2.2 miles recalled by Picuris consultants for the turn of the century. The *acequia madre* is actually a u-shaped canal three feet in width and depth. A log and boulder diversion dam spans the Rio Pueblo in the canyon east of the pueblo and provides a flood pool from which a head gate diverts water into the canal. The canal, which is a gravity flow system, joins the river again west of the village and its associated farm land. Lateral ditches and smaller feeders in each field distribute the water from the ditch as needed. The upkeep of the canal is a community responsibility under the direction of civil officers. The smaller ditches, however, are owned and maintained by individual farmers and are sometimes abandoned.

Naturally the canal provides water and new soil to the fields, but it has other important purposes for the Picuris as well. Ditch water is drinking water, bath water for youngsters, and wash water for clothes and wheat. Occasionally, with prior permission from the ditch boss, young men close the ditch and in an hour's time women and children are scrambling for the tasty trout stranded along its bottom. Simultaneously the *mayordomo* quickly inspects the ditch, removing obstructions and making emergency repairs. Moreover, the bank is a man-made plant community of extreme value to the community. Table 1 summarizes these foods which are free to everyone. Some are perennial and others are annuals favoring the backpile following the spring clean out. The canal is obviously intrinsic to the economy.

Since the irrigation hypothesis centers on the community organization required to clean the ditch and on the coordination of distribution of water, these two issues will be treated separately.

Cleaning the Canal

The annual spring cleaning of the ditch is labor intensive and very arduous. The supporters of the irrigation hypothesis argue that many would shirk the toil, that they must be coerced to participate, and that the overall effort requires a special centralized management not found in other aspects of Pueblo life. Data from Picuris argue otherwise.

Each year before planting begins the ditch must be repaired. Accumulations of debris from the previous year have filled the bottom unevenly. Flood waters have carried large stones into the canal and the sequence of winter freezings and thawing encourages the bank to slump. Other damage results from burrowing animals, an occasional beaver making a home in the bank, or domestic stock caving in the sides. This damage must be repaired to prevent water loss and to assure a continuous water supply through the summer.

With the melting of the snow the ditch boss examines the canal to determine the time for its annual cleaning. The ditch must be ice-free and dry. One year it may be ready in mid-March and the next not until late April.[5] For five days prior to the date set by the *mayordomo* and the governor, the War Captain announces the event. At dawn on the appointed day, the men and older boys gather on the north side of the village in front of the cacique's home and sing in a circle until all are assembled. The cacique will pray during the day, and will conduct the ceremony to open the ditch when the work is concluded. They then move to the ditch and start cleaning from the middle of the village to the south where it joins the river.[6] Each uses his shovel or digging stick to straighten the sides of the ditch and clean the bottom of all silt, stones, and debris. At

TABLE 1 Useful Plants Growing on the Canal Bank

Season	*Name*	*Usage*
Early Spring	wild potato *(Solanum jamesii)*	eat tuber raw or cooked
Spring	cutleaf coneflower *(Rudbeckia laciniata)*	young leaves for spinach
	Mountain parsley *(Pseudocymopterus montanus)*	flowers for condiment
	lambs' quarters *(Chenopodium album)*	young plants for spinach
	wild mint *(Mentha arvensis)*	medicine
Summer	wild rose petals *(Rosa woodsii)*	sore throats
	milkweed *(Asclepias latifolia)*	leaves for spinach; young pods for snack
	yarrow leaves *(Achillea lanulosa)*	condiment; fever medicine
	horsetail *(Equisetum laevigatum)*	stem for whistle
	grass	for brooms
Fall	wild plum fruit *(Prunus americana)*	eat fresh; store dried
	rose hips	eat fresh; store dried
	dogbane *(Apocynum cannabinum)*	bark for string; stem for toy
Winter	Wild plum bark	boil for cough

mid-day the men sing a series of special ditch cleaning songs while the women and children bring food. A communal picnic ensues with the men sitting apart from the women. The meal is punctuated by amusement and further singing. Following the picnic young boys hold informal races in preparation for several ceremonial races which will follow during the planting season. In the afternoon the women may be asked to assist by carrying stones or dirt to repair serious breaches in the canal bank. By sundown some 12 hours of hard labor have passed and better than a mile of ditch has been cleaned.

On the second day the same procedure is followed except that this time they work from the village to the head gate. At the end of the second day the ditch is ready to receive water. Following prayers by the village cacique and the ritual planting of turkey feathers in the ditch by the men and women, the water is permitted to flow through the canal. Cleaning the canal is part of a ritual act to bring blessings to the people.

While the afternoon contribution of the several women who assist the men is important, for these two days the basic work team consists of some 30 men and boys. Old or infirm men patrol the bank offering advice, giving encouragement, and leading songs or telling stories about how much harder it was in the "old days." No recognizable authority figure polices the work. The ditch boss, who is chosen because of his farming prowess, approves the completion of each man's trench, but he will become just another worker the next year when someone else is appointed *akissawin.* Within the Pueblo this two day activity is universally acknowledged as one of hard work—each worker has cleaned almost 400 feet of ditch—and is hailed as a good time.

During the year the *mayordomo* will adjust the head gate, make minor repairs, or ask the Council for assistance when the ditch is endangered. Usually the ditch boss and his assistant can handle most contingencies.

A minimum of direction is required and in fact the Picuris themselves have instituted a *mechanical* solution to the social problems suggested by academic theoreticians. When a recent governor was asked about any coercion or recalcitrance, he replied:

> The Old Ones who built the ditch were pretty smart. They knew that if they started cleaning from the East (head gate), as they passed a man's field he would drop out from the work. This way (cleaning toward the head gate) no one gets water until the (entire) ditch is clean.

Ditch cleaning is only one of several community endeavors, and certainly not the most complex or the most highly orchestrated. Each year a group of Picuris families go through the Mora Valley to the high plains of eastern New Mexico to hunt bison in a manner similar to Southern Plains nomads. At other times the Hunt Headman organizes all the men into a late winter deer drive. Hunters are stationed at the tops of the ridges in the steep valleys north of the village to ambush deer driven by beaters moving up from below. In a previous era defense and counter offense against marauders were occasionally required under the direction of the War Captain. Ritual performances, however, may require the greatest coordination of the entire village over the longest period of time. For one example, the rehearsal of songs and dances for San Lorenzo day celebrated on August 10th begins twelve days prior to the actual festivities which include a Vespers Dance on the 9th, a Racers' Dance the morning of the 10th, a spirited footrace between the North Side and South Side groups, another community dance in the afternoon, and a burlesque and pole climb by the Water Clown Society. While these public activities are presented, friends and visitors are feted in private homes, and members of the Council of Principles are in prayerful retreat in the kivas. By comparison ditch cleaning is a picnic!

Availability of Water

The ditch boss, the governor, and even the Council are prepared to hear the petition of an aggrieved farmer who is receiving insufficient water. Do such conflicts over the allocation of water ever arise, and if so, how frequently?

To answer this question we must return to our study of the physical aspects of technology and examine the crops, size and distribution of irrigation fields, the amount of available water, and the frequency of use. Each of these variables is used in computing the amount of irrigation water required (Blaney and Hanson 1965).

No authority in Picuris allots water; no calendar is followed. A farmer's perception of the condition of his crops determines when he will irrigate. The climate in this mountain valley limits the variety of plants that can be cultivated. Cold hardy cultigens, such as peas, onions, and potatoes, are grown in kitchen gardens tended by both men and women. The field crops are limited to Spanish-introduced wheat the aboriginal corn, beans, and squash. More land is devoted to production of corn than of anything else, and since it has the greatest water requirements, it determines the operation of the irrigation system.[7]

The importance of corn in the Picuris way of life also determines the distribution of fields. Although inheritance patterns lead to households having scattered holdings of much less than an acre, no attempt is made to increase efficiency by consolidation. Corn is classified according to seven different colors with five of these coinciding with the five colors symbolic of the cardinal directions. These colors must be raised separately and kept pure for ritual and dietary reasons, which would be an impossibility if all colors were grown in the same field.[8]

Pure color corn is a necessary component in various ceremonies and its maintenance is reinforced in other rituals. East is associated with white corn, north with black, west with yellow, south with blue, and the "in the middle" or southeast with gray; corn is planted in the same order. Long ago, perhaps in conjunction with the Spring Picnic, an annual five day ritual under the sponsorship of the cacique and his society (Brown 1973:103), a corn dance for each of these was held on separate days. Throughout the year there are rituals requiring baskets full of ground corn of a particular color in order to be efficacious.

Corn is a staple of the diet, with each dish defined according to the appropriate color of corn. Basically, white cornmeal is consumed as porridge, paperbread, chicos, and posole. Blue cornmeal is preferred for atole and a special piki, although gray and red are prepared in this manner as well. Yellow is for corn husk baked corn bread. Black is generally fed to stock.

Each of the 30 families plants between three and five acres of land divided into five to eight fields. The total land planted barely exceeds 100 acres (Walpole 1900:293). All parcels are adjacent to the canal or are connected with it by a lateral ditch. The fertility, texture, and composition of the soil does vary but overall everyone has access to every type. The same applies to the contour of the fields. An even surface increases irrigation efficiency but no field is perfectly flat and no farmer is saddled with land consistently more uneven than that of others.

The amount of water required by corn has been computed on a monthly basis in Table 2. It assumes that only 55 percent of the irrigation water reaching the field is utilized, and, as a result of field contour and loss to feeder ditches or overflow, another 15 percent must be added to that entering each field if all plants are to obtain an equal amount of moisture. Picuris farmers use a furrow method for distributing water between the rows. This technique requires the full attention of a farmer while the canal is open; thus only one field can be irrigated at a time. Each field is irrigated only three to five times depending upon the moisture held by the soil at planting time and the amount of rainfall during the growing season. Water is turned into the field until it stands about 4 inches deep. Since planting times are not synchronized, the number of acres planted varies according to domestic and ritual obligations, and microenvironmental factors affect rates of growth, each farmer determines his own irrigation schedule.

To better understand the possibility of conflict over water ever arising we can examine the amount of water available under the most severe conditions. Table 3 summarizes published acre feet of water in the Picuris ditch and the Rio Pueblo. We have already indicated that the effective moisture is lowest during the month of June when irrigation water is imperative. If we assume that there is a very dry spring, it might be necessary for each of 30 farmers to irrigate each field twice. Referring once again to Table 2, the water required for corn in June is .60 acre feet. To

TABLE 2 Model of Monthly Irrigation Water Required by Corn Grown at Picuris Pueblo

Month	*[1]Consumptive Use*			*[2]Effective Rain-fall (r)*	*Consumptive Use Minus Effective Rainfall (u−r)*	*Field Irrigation Requirement (I)*		
	Factor (f)	*Coeffi-cient (k)*	*Amount (u=fk)*			*+55% efficiency*	*+15% head gate efficiency*	*Total*
			Inches	Inches	Inches	Inches	Inches	Acre Feet
May	5.38	.50	2.69	1.3	1.39	2.53	2.91	.24
June	6.28	.70	4.40	.94	3.46	6.29	7.23	.60
July	6.83	.80	5.46	1.8	3.66	6.65	7.65	.64
August	6.33	.80	5.06	1.9	3.16	5.75	6.61	.55
September	5.03	.70	3.52	1.3	2.22	4.04	4.05	.39

Sources: [1]Blaney and Hanson 1965
[2]U.S.D.C. Climatological Data: Peñasco Ranger Station

irrigate all the land twice at the maximum figure would thus require 120 acre feet per month. Since the least amount of water carried by the ditch is 162 acre feet or 5.4 acre feet per day, there is sufficient water to meet everyone's needs.

Conflict can occur when too many farmers attempt to irrigate on the same day. Even under the conditions outlined above, 6 farmers could irrigate at least 6 acres (or two fields each) per day assuming the ditch is used for 6 hours which it often is. If, as is quite possible, more than 6 farmers wish to irrigate, disappointment is inevitable. The usual resolution is that the disgruntled farmer wakes up the next day before dawn and irrigates without competition.[9]

The conditions we described as necessary to create conflict rarely occur. The amount of water is more than sufficient and not all the fields are planted in corn. Furthermore, the average amount of water in the ditch (Table 3:288 A.F.) is ample, even in a dry year, to allow half the farmers to irrigate almost an acre each day. Disagreements do arise between neighbors but not over water.

DISCUSSION

Wittfogel (1957:88) recognizes the limits to power in the hands of Pueblo ceremonial officials and the incipient centralized authority of these hydraulic tribes. Nevertheless, his assessment and the conclusions of those who accept the irrigation hypothesis have failed to provide a technological understanding of irrigation in the Eastern Pueblos.

Picuris is not an exception. The same attitudes about ditch cleaning are found in San Juan and Cochiti Pueblos (Ford, field notes). These subsistence farmers realize that the canal is their lifeline and that it would be foolhardy to jeopardize their livelihood or their technological base by failing to repair and to protect the canal. No authority is needed to enforce this lesson.

By viewing irrigation as a technological problem, we have enhanced our understanding of its social consequences. The assumptions underlying the irrigation hypothesis are not supported by the Picuris study. And, in fact, any officer who has some role in the management of the hydraulic system is prevented from assuming further authority by being replaced on a yearly schedule. Further study of the technology shows that reluctance over the cleaning of the canal is not resolved by force but simply by cleaning in an upstream direction and that the likelihood of conflict over the operation of the irrigation system is remote. Irrigation is not the prime mover of Pueblo social organization.

TABLE 3 Picuris Water Supply 1936–1941

Source		*Mean and Range per month (in Acre Feet)*											
		Jan.	*Feb.*	*Mar.*	*April*	*May*	*June*	*July*	*Aug.*	*Sept.*	*Oct.*	*Nov.*	*Dec.*
Rio Pueblo (500' below head gate)	Mean	612	705	1744	7632	15972	6483	1998	1153	949	916	796	712
	Low	389	606	962	4320	6670	1510	168	391	490	631	607	595
	High	793	897	2520	13460	45990	45990	6220	4020	1680	1480	1080	801
Picuris ditch (300' from head gate)	Mean	0	0	0	14	88	288	324	262	137	29	3.3	0
	Low	0	0	0	0	4.8	162	256	138	54	6.5	0	0
	High	0	0	0	34	166	423	471	352	214	54	4.6	0

Source: New Mexico State Engineer (1959:220)

NOTES

[1]Based on the usual inflated Spanish estimates, Picuris is said to have contained as many as 3,000 inhabitants prior to the Pueblo Revolt of 1680. However, using the figures given in this paper as a guide even a population one-fourth that size would have taxed the hunting and farming potential of the region.

[2]Climatic figures from Taos are used because of their continuous record of reliability. Since Taos is several hundred feet lower and in a more open location, its temperature is higher and precipitation lower. Rainfall and yearly precipitation figures are from Peñasco Ranger Station, located 2 miles east of the village.

[3]The figures are derived from Blaney and Hanson (1965). Monthly consumptive-use (u) is the evapotranspiration rate of water loss from the soil and from plants.
$u=kf$ where,

$t=$ mean monthly temperature

$p=$ monthly percent of total daytime hours of the year

$f=\frac{t \times p}{100}=$ monthly consumptive-use factor

$k=$ is a measured coefficient for each irrigated crop grown under conditions (f)

$u=$ monthly consumptive use of water by plants in inches

[4]*Mayordomo* is a Spanish term meaning steward or manager. The English colloquial term is ditch boss. The Picuris term is *akissawin* which is obviously derived from the Spanish word *acequia,* an irrigation canal or ditch.

[5]On April 1, 1975, several inches of ice remained along the bottom of the canal.

[6]One consultant recalls that "long ago" the ditch was cleaned starting from the end and concluding at the head gate. This was not confirmed by others, but structurally it is compatible with the conclusion of this section.

[7]Corn is limiting because its seasonal consumptive-use coefficient (k) is higher than those of the other plants. See Blaney and Hanson (1965:22–23) for details.

[8]The separation of fields according to ritual and ceremonial percepts has a latent ecological consequence not mentioned by consultants. Because of localized environmental vicissitudes an entire crop could be lost to frost, hail, or grasshoppers if it were planted in one large field. As it is, some loss occurs but the dispersal of fields mitigates against ecological disaster.

[9]Irrigation agriculture is very labor intensive but not for the upkeep of the system. The water puddles the soil and fills the pores with silt. This must be broken up to increase surface area for absorption and to permit gas exchange at the roots. Thus several days of hoeing must follow each irrigation.

LITERATURE CITED

Blaney, Harry F. and Eldon G. Hanson

1965 Consumptive Use and Water Requirements in New Mexico. New Mexico State Engineer Technical Report No. 32.

Brown, Donald Nelson

1973 Structural Change at Picuris Pueblo, New Mexico. Unpublished Ph.D. dissertation, University of Arizona.

Donaldson, Thomas

1893 Pueblos of New Mexico, *In* Moqui Pueblo Indians of Arizona and Pueblo Indians of New Mexico. Thomas Donaldson, ed. Pp. 1–49. Eleventh Census of the United States, Extra Census Bulletin.

Dozier, Edward P.

1970 The Pueblo Indians of North America. New York: Holt, Rinehart, and Winston.

1960 The Pueblos of the South-Western United States. The Journal of the Royal Anthropological Institute of Great Britain and Ireland 90:146–160.

Eggan, Fred

1966 The American Indian. Chicago: Aldine.

Müller, Frederick P.

1893 Picuris *In* Moqui Pueblo Indians of Arizona and Pueblo Indians of New Mexico. Thomas Donaldson, ed. Pp. 118–119. Eleventh Census of the United States, Extra Census Bulletin.

New Mexico State Engineer

1959 Hydrologic Summary. New Mexico Streamflow and Reservoir Content 1888–1954. State of New Mexico, State Engineer Office Technical Report No. 7.

Service, Elman R.

1975 Origins of the State and Civilization. New York: W. W. Norton.

Vivian, R. Gwinn

1974 Conservation and Diversion: Water Control Systems in the Anasazi Southwest. *In* Irrigation's Impact on Society. Theodore E. Downing and McGuire Gibson, eds. Pp. 95–112. Anthropological Papers of the University of Arizona 24.

Walpole, N. A.
1900 Report of Agent for Pueblo and Jicarilla Agency. Pp. 292–297. Annual Reports of the Department of the Interior for the Fiscal Year Ended June 30, 1900. Indian Affairs.

Wittfogel, Karl A.
1957 Oriental Depotism: A Comparative Study of Total Power. New Haven, Conn.: Yale University Press.

Wittfogel, Karl A., and Esther S. Goldfrank
1943 Some Aspects of Pueblo Mythology and Society. Journal of American Folklore 56:17–30.

Wright, Henry T.
1976 In press Toward an Explanation of the Origin of the State. *In* Explanation of Prehistoric Organizational Change. James N. Hill, ed. Albuquerque N.M.: University of New Mexico Press.

7

The Energy Costs of Technology in a Changing Environment: a Machiguenga Case

ALLEN JOHNSON

University of California, Los Angeles, California

INTRODUCTION

In this paper I examine the process by which a community of hunter-collector/horticulturalists of the Amazon rain forest is increasing its dietary dependence on domesticated foods at the expense of naturally-occurring foods. The data will show that the energy costs of production of dietary protein are higher under the hunting-and-collecting regime than under domesticated food-production. The change that is occuring, therefore, is in the direction of increased energy efficiency, or so it seems. But the interpretation of this result is not at all simple, and a goal of this paper is to indicate some conceptual difficulties with "energy-efficiency" as a variable to explain technological change.

We tend to assume that there is an economy in living systems such that energy is not allocated within and between organisms unnecessarily or inefficiently. It may be that this cannot be proved, or even tested, but the fact of intra- and inter-specific competition for limited food resources makes it highly likely. By extension, human beings, as biological organisms, should elect practices which accomplish their goals with a minimum of waste effort. Link this idea to the idea that technological change throughout history has served to *increase* the efficiency with which basic human needs for shelter, food, and defense have been met, and the basic structure of an influential and plausible theory of cultural evolution emerges.

In the development of culture-evolution theory from, for example, Childe (1951) to White (1959) to Harris (1975), we find this common thread (substantial improvements in the clarity of the argument notwithstanding): that the general trend of food-producing technology through history has been toward a definite, large increase in food production per unit of human labor input.

The matter has been confounded, however, by an equal and opposite opinion. Boserup (1965), aruging as an agricultural economist and not as a culture-evolutionist, offers this reasoning (I have generalized it from agriculture to all food-getting activities): within a given land area human beings are presented with a range of subsistence options, some of which require less work than others; as population numbers within the area increase, the "easy" options become exhausted and people must work more to maintain the same subsistence level. Technological change in many cases is a process of intensification, therefore, in which the number of people who can live on a given area of land is increased, but at the expense of leisure time. Accordingly, the efficiency of food-producing technology *declines.*

Each argument no doubt has its range of application, so that neither alone can adequately describe the course of technological change in history. Here I am more interested in what the arguments have in common and take for granted: that there is such a thing as the efficiency of a technological system as a whole and that efficiency in this sense is conceptually necessary for the theory of technological change.

In this paper I will describe a case of technological change in terms of "efficiency," operationally defined in this case as the energy costs of producing a given good or service. While I hope it will become clear that expressing technological changes in terms of energy efficiency lends a strong, concrete dimension of understanding, the results of the analysis are not very usefully described as the replacement of an inherently less efficient technology by an inherently more efficient one, or vice versa.

The analysis will indicate that the efficiency of specific techniques may itself change gradually in sensitive adjustment to other changes in the ecosystem. Whether the efficiency of the technological system as a whole has changed remains in doubt, and is perhaps unanswerable.

MACHIGUENGA PROTEIN PRODUCTION

In general we would like to know whether and under what conditions technological systems change in the direction of greater labor efficiency. An approach to this problem is to view a technological system in the process of change to learn how alternative techniques compare in terms of labor efficiency. The case examined here is a shift in the techniques of producing dietary protein among the Machiguenga Indians of the Upper Amazon of Southeastern Peru. The Machiguenga live in small groups (7 to 25 members) scattered throughout the rain forest of the Urubamba watershed. The community discussed here (Shimaa) is at an altitude of 2300 feet on the Eastern slopes of the Andes, where game and fish are less abundant than in the lower altitudes. They are almost completely self-sufficient producers of the foods and manufactures they use, and enjoy good health, apart from parasites and a dangerous susceptibility to colds and influenza. In the past few years they have begun to decrease their dependence on protein foods hunted and collected in the forests and rivers in favor of protein-rich garden foods and domesticated animals. The change is directly related to recent increases in community size and access to trade goods which have followed the establishment of Peruvian schools run by Machiguenga schoolteachers. We may first review the sources of Machiguenga dietary protein, and then evaluate current changes in the energetic efficiency of each technique.

Sources of Protein

The four main sources of protein in the Machiguenga diet are hunting and collecting in the forest, fishing and collecting small fauna from local streams, growing maize, and raising domesticated animals (primarily chickens and ducks). Data presented in this section describe the "labor efficiency," or "energy efficiency" of each of the protein sources.

The measure of "efficiency" being used here is the rate of return of protein per unit of energy expended in protein-getting activities. The data were collected in a number of ways. First, a random sample of daily activities was taken throughout the first ten months of residence in the vicinity of the school at Shimaa, along a tributary of the upper Urubamba River. Nearly 3,500 random observations of individuals were made, fairly representing times of day, season, and individuals of both sexes and all ages; the resulting statistical patterns are reliable indicators of the overall pattern of time expenditure in the population (Johnson 1975). Estimates of total amounts of time spent in protein-getting activities are based on these observations.

Estimates of the amounts of energy expended come from measurements made by Dr. Edward Montgomery and myself using standard calorimetry procedures (Montgomery and Johnson 1975). Rates of energy expenditure were measured for all the major tasks performed by Machiguenga men and women. Rates of energy expended in protein-getting activities, multiplied by the amounts of time spent, give the total amount of energy in calories expended in the various protein-getting activities.

Direct observations of a number of hunting, collecting, and fishing expeditions provide data on the amounts of wild foods yielded by these activities. Measurements were also made of garden sizes, crop densities, and amounts of weight of harvested foods; a census of household fowl and observation of breeding practices round out the quantitative data.

The calculation of energy efficiency measures at this point would appear to be straightforward, but in fact it is not. The problem arises in defining which activities are regarded as necessary for the production of protein foods. This is part of the more general problem anthropologists have in defining what they mean by "productive activities" (Johnson 1975:302). It is clear that anthropologists are far from agreeing on what activities are "food-producing activities." Sahlins (1972), in comparing amounts of work expended in food-production cross-culturally, is forced to compare data on New Guinea horticulturalists which include only garden labor with data on Australian Aborigines which count all food-related activities, including cooking and tool-manufacture as well. Such serious disparities of definition clearly restrict the usefulness of cross-cultural comparisons of this sort.

To illustrate the difficulty in the present case we may try to decide which activities contribute to protein production directly, which contribute indirectly, and which contribute not at all. Obviously, all effort expended directly at the site of protein production (forest, river, maize garden) is relevant (exempting small exceptions, as when couples go to the forest not only for food, but also for privacy and love-making). Certain other activities appear unambiguously to contribute to protein production as well, such as bow and arrow manufacture or labor traded for a steel axe.

But, how are we to count the labor paid for the axe? Axes are used to dig grubs from trees, but they are also used in many other activities. One solution might be to estimate the proportion of axe use which is devoted to removing grubs from trees, and allocate that proportion of the labor to the energy costs of protein. This solution, although difficult from a practical standpoint, is still clear enough conceptually; some part of the energy cost of the axe is an indirect cost, at one step removed, of getting protein. Axes are also used, however, in the manufacture of bows and arrows, which in turn are used to obtain protein—should this indirect cost, now two steps removed, also be included in the energy cost of protein? At this point, serious conceptual difficulties arise (and the practical difficulties go right through the roof). Men generally use axes to obtain large pieces of hard palmwood, which they store in the rafters of the house. Some of this wood is used to make bows and arrows. But the wood may also be used to make the spindles which the women use to spin cotton thread, small amounts of which the men then use to manufacture traps; the chain of indirect costs is getting longer. Then, too, the palmwood is used to manufacture wood nails for balsa rafts and toy spinning tops for children—someone with imagination could probably connect these costs to protein production as well. What procedure is there, acceptable to a majority of anthropologists, each with experience in different cultural settings, for deciding which indirect costs should be included, and which excluded, from the calculation of protein costs?

The expedient I have adopted here is to count only the direct labor inputs at the site of protein production. Although labor expended in the forest produces more than just protein, every trip to the forest is an attempt to find protein, and therefore is counted as protein-productive. By contrast, many gardens which do not contain maize do not produce significant protein, so only labor invested in maize gardens is counted as protein-productive. Support activities, such as tool manufacture and food preparation, are excluded for the moment, because they raise the definitional problem just mentioned.

Hunting and collecting in the forest takes about 9 percent of men's and 2 to 3 percent of women's daylight time (Johnson 1975:308). Only men hunt, usually alone or in pairs, for 4 to 6 hour periods once or twice a week. Women often accompany their husbands, in order to help collect foods in season, including palm hearts, fruits, nuts, and grubs and other insects. When game is sighted, men become silent and stalk the prey with bow and arrow; the women, if they are along, wait beside the trail. The most commonly hunted game are wildfowl (guans, currasows, toucans), spider monkeys, and peccaries. Large game, even the peccaries, are rare, and tapirs, deer, and other mammals are almost never seen.

Indeed, the forest offers a low rate of return to labor. Labor costs are high because the forest is mountainous, trails are steep, and much exertion is necessary to cover a broad enough extent of forest (the highest observed rates of energy expenditure were during forest walks). In addition, all foodstuffs are widely scattered in the forest, and more often than not hunters return emptyhanded or with small amounts of collected foods.

Table 1 shows the low returns the Machiguenga obtain from labor invested in edible forest products. Column 1 lists the categories of foods obtained from forest expeditions; Column 2 gives the average amounts of food, in grams, which are brought in for every *kilocalorie* (the conventional unit in which energy-expenditure in work is expressed) of energy spent in the forest; Column 3 gives the energy-value in Calories (the conventional unit for expressing the energy-producing value of food when oxidized in the body) for every kilocalorie expended (one kilocalorie equals one Calorie, despite the conventional difference in terms); and Column 4 provides the estimates of the protein content in grams per unit of expended energy.

The Calorie and protein values of foods are based on published analyses (Bodenheimer 1951; INCAP 1961; Latham 1965; Platt 1962). Since not all types of Machiguenga foods are treated in published analyses (e.g., larvae of local insect species), and since the nutrient content of even well-known foods varies from place to place, the figures are necessarily approximate. Future refinements may lead to small changes in the specific figures given below, but the overall pattern of differences seen in Table 3 is not likely to change significantly.

TABLE 1 Energy Costs of Forest Foods

(1) *PRODUCT*	*(2)* *g./kcal.*	*(3)* *Cal./kcal.*	*(4)* *g. protein/kcal.*
Game	.138 g.	0.16 Cal.	0.017 g.
Grubs, etc.	.106	0.11	0.032
Palm Hearts	.356	0.12	0.009
Fruits	.534	0.32	0.005
Nuts	.023	0.13	0.003
TOTALS		0.84 Cal.	0.066 g.

The forest compensates for poor protein returns to some extent, in two ways. First, the forest provides a variegated supply of foodstuffs which cannot be reduced to a simple calculation of calorie or protein values. Much of the qualitative richness of the Machiguenga diet derives from forest products, which provide vitamins and other substances which might not otherwise be available. And second, it is the forest that provides the raw materials upon which their whole technology rests: fibers and hardwoods used in houses, tools, boxes, nets and other necessary items of manufacture. Frequent, wide-ranging excursions into the forest are valued as ways of locating these resources and reporting their whereabouts to relatives and neighbors. A declining reliance on the forest as a protein source, therefore, has complex implications for overall dietary balance and technological self-sufficiency as well.

Fishing is for the Machiguenga a substantially more effective method of obtaining protein; it occupies men 6 percent and women 2 percent of the daylight hours. The major fishing techniques are fish poisoning, fishing with nets, and casting with hook and line. Poison (barbasco) fishing is the most productive; it often involves as many as 6 to 8 households, a degree of cooperation not found in other Machiguenga activities. Separate dams are constructed upstream, by the men, and downstream, by the women, and then the slowed waters in between are saturated with the fish poison, which is obtained from the roots of one or more cultivated plants. The method, however, is so effective that it virtually eliminates the population of fish and other water fauna for the stretch of river between the dams, and since there are only a few suitable places for such fishing, it is not repeated until a period of time has passed. During our fieldwork we observed only four cases of barbasco fishing. Fishing with nets is done by men, using a large net, and women, with a small net, when the river is muddy from rain. Nets are shovelled along the bottoms of rocky streams to dislodge small fish and other fauna ranging in length from 1 or 2 inches to perhaps 6 inches.

Hook and line fishing is practiced only about one-third as often as either of the other two techniques, and is the least productive. Fishermen will cast a line in the morning or evening when they go to the stream to bathe, but generally without success. I have often seen men return after several hours to find the bait, usually a grub, still untouched on the hook. When fish are caught, however, they are large for this region, weighing from two to five pounds.

Table 2 presents the fishing returns per unit of energy expended. Columns (2) and (3) happen to correspond in this case because one gram of fish yields approximately one Calorie of food energy. The weighted averages for Columns (3) and (4) were calculated by multiplying each value by the decimal proportion of total fishing time which was spent in that particular fishing activity, and then summing the results; thus the weighted average is close to the actual efficiency of fishing as a whole, according to frequencies of each type of fishing on a yearly basis.

Other fishing techniques, such as weirs or bow and arrow, were not observed in this community, although one man possessed an iron-tipped arrow he said was specifically for fishing. Dynamite is used occasionally, when available (it is illegal), but has the same devastating effect as barbasco, and must be seen as a replacement technique, rather than as a new technique which would increase overall production. Dynamite may be used without dams, but then many of the fish killed by the explosion float away downstream before people can get to them.

TABLE 2 Energy Costs of Aquatic Foods

(1) *TYPE*	*(2)* *g./kcal.*	*(3)* *Cal./kcal.*	*(4)* *g. protein/kcal.*
Barbasco	2.82 g.	2.82 Cal.	0.56 g.
Nets	1.22	1.22	0.24
Hook & line	0.65	0.65	0.13
WEIGHTED AVERAGES		1.95 Cal.	0.39 g.

Livestock raising has long been an available protein-producing option for the Machiguenga, in the form of the Muscovy duck, an aboriginal domesticate. Nonetheless, domesticated animals appear to have made a very small contribution to the overall diet in the past, before the stable settlements of the past few years came into being. Now chickens as well as ducks are being raised in increasing numbers. Aside from initial construction of a house for the fowl, they require only a few minutes of labor in care each day. Chickens forage for seeds and small insects in the environs of the house, and for the most part feed themselves. An ear or two of maize, or some bits of cooked manioc, are the only garden contributions to their diet. Their returns to labor are 0.70 grams of protein and 9.28 Calories per kilocalorie, substantially better than either forest or aquatic products.

But a low number is imposed on the household flock by the predations of wildcats and by the near absence of natural feed in the tropical forest environment; the same environmental deficiencies that limit game populations limit the foraging efforts of domestic fowl also. Any attempt to increase flock size by protecting the fowl from predators would certainly raise the labor costs, but the problem of feed is the more critical one. Chickens, for example, when they do not feed themselves, must be fed a diet which contains at least 15 percent protein, as well as calcium, and many other nutrients (FAO 1965). This would mean diverting protein-rich foods like maize or fish from humans to the household fowl, and since these foods require human labor, this would reduce the labor efficiency of livestock protein.

Finally, gardens provide potentially large and energy-efficient sources of protein for the Machiguenga diet. Garden labor occupies the men over 18 percent of their daylight hours, and the women nearly 7 percent. At present, the only protein-rich food grown in any quantity in Machiguenga gardens is maize. Beans, peanuts, and peas are known, but their actual frequencies in gardens are low; all three foods are regarded as tasty, however, at least in small quantities.

Maize is planted only in first-year gardens, along with manioc and a variety of other root crops. In first-year gardens maize accounts for nearly 54 percent of all plants; I will use this figure as an estimate of the proportion of total labor in first-year gardens which is devoted to maize cultivation. In gross terms, maize cultivation is the most efficient method of producing protein; the figures are, 1.18 grams of protein per kcal. of labor input, and 45.40 Calories per kcal.

The problem with maize as a protein source is well known, however; maize provides incomplete proteins which alone cannot sustain human life (FAO 1964:8–14). Since the Machiguenga know of beans, it is interesting that they do not grow them. In fact, in the community of Mantaro, some days travel by footpath from the one in which the present study was done, at a higher altitude where game and fish are even more scarce, the Machiguenga rely much more heavily on beans. Perhaps beans are an unattractive option used only when other sources become unreliable.

Rice was introduced to the community by way of local Peruvian horticulturalists during the year of this study (1972), and was planted experimentally by two Machiguenga men. During a recent visit (1975) I found several more men planting rice regularly, and a few experimental stands being tested in housegardens by the more cautious ones. Since rice is also a source of high quality protein, and is enjoyed as a food by the Machiguenga, and since it is planted in addition to maize, not as a replacement, it may well become an important additional source of protein in the years to come.

Energy Costs

Table 3 presents the comparative figures for energy costs of calories and proteins under the four methods of hunting, fishing, poultry raising, and maize cultivation. Each of the four activities can be readily ranked relative to the others. The differences are so great that we would predict, from the standpoint of labor efficiency alone, that current change would be in the direction of greater reliance on poultry and maize, and the abandonment of reliance on forest and river sources. Indeed, this is what is happening. But, since the maize-poultry option has been available for years and probably centuries, we need to consider why the predicted technological transformation did not take place long ago, and why, even now, it is progressing slowly.

DISCUSSION

To what extent are the observed changes in Machiguenga food-getting techniques an internal response to the simple attractions of labor efficiency, and to what extent are exogenous influences bringing the change about? We must consider two kinds of exogenous influence: first, the easy access to steel tools in recent years; and, second, attempts by the Peruvian government to increase the size and permanence of Machiguenga settlements. These two processes are not separate but rather closely interconnected.

We have no good data on the point at which steel tools were introduced to the Machiguenga. Informants report, however, that until the last 10 years or so, they had difficulty obtaining steel axes or machetes, these being known but available in small numbers and hand-me-down condition in isolated forest settlements. Informants report that then their gardens were much smaller than today and they depended heavily on the forest for food. Old stone axes are turned up in gardens from time to time, but present-day Machiguenga have no knowledge of stone axes, and have almost certainly used whatever fragments of steel they could get for garden labor for the last several generations.

If we assume that the present abundance of steel tools of the right size in good repair has doubled the efficiency of *clearing* a garden (which now accounts for 33 percent of labor in maize), the cost of maize protein would have been about one-third higher in the past, and this would reduce maize protein, considering its low biological value, to being perhaps no better a source than fishing.

Ease of access to steel tools is not without its political aspect. The Peruvian nation is seeking to extend the scope and intensity of its influence over the portion of the Amazon which is within its territorial boundaries, both in order to ensure its borders with Brazil, Ecuador, and Bolivia, and to increase the value of the region as an exporter of goods and as a tax base. This has meant the provision in recent years of educational

TABLE 3 Summary of Labor Costs of Protein Production

Source	*Cals./kcal.*	*g. Protein/kcal.*
Forest	0.84 Cal./kcal.	0.07 g./kcal.
Rivers	1.95	0.39
Poultry	9.28	0.70
Maize	45.40	1.18

facilities, medical care, and not least of all, tools, in order to attract settlers scattered throughout the forest into large, sedentary communities which can be directly subject to Peruvian law, and which will provide more market opportunities.

In the area where the present study was done, this has meant that over 125 individuals have settled within an hour walking distance of the new schoolhouse, with the result that the effective population density within the school vicinity has increased several times above the traditional level (at the same time, of course, distant regions are depopulated, but, being distant, their resources are unavailable to members of the school community).

We have already seen that the forests of the Machiguenga region are too poor in resources to support a much greater population—as it is, hunters spend long hours for relatively meager returns. The Machiguenga believe, and observations in areas far from the school community confirm, that game is much more readily encountered around the isolated forest settlements than is currently found in the vicinity of the school community. Hunters nowadays expect a one to two hour hard march up the forest slopes before seriously beginning to look for game. The newly increasing population densities under which Machiguenga are beginning to live contribute to a lowering of the labor efficiency of obtaining protein from wild resources. A similar argument applies to fish.

Thus, the present labor-efficiency of food-getting activities has certainly been influenced by recent external changes which have tended to raise the efficiency of domesticated sources of protein relative to the wild sources. But even with these factors in mind, it hardly seems likely that in the recent past the *forest* in the Machiguenga area was ever a relatively efficient source of protein. Why was the forest not abandoned long ago, when the first agriculture became available? The answer lies in the qualitative diversity of the forest as a source of Machiguenga provisions. Each "hunting" trip produces not only a certain amount of protein-rich food, but also provides a wide range of foodstuffs, of rich and complex nutritive content. Additionally, such trips provide a continuous flow of information concerning the whereabouts of raw materials and of out-of-season foodstuffs which will become available later.

Under their new settlement conditions the Machiguenga are finding all forest products ever more scarce. For example, the members of the "school community" have used up all the best roofing leaves within a two-hour walk, and the most recently-built houses have required long expeditions to get the material, or else the substitution of materials unanimously regarded as much inferior. The same is true of the hardest woods for house mainposts, and the particular palmwood preferred for walls and sleeping platforms. Likewise, individuals who locate ripe fruits or large numbers of grubs and return with their families to collect them are now liable to find that in the meantime someone else has taken them.

SUMMARY

The process of change in the Machiguenga technology of protein production is obviously complex. Here I have only been able to suggest its outlines. The data show that domesticated sources of protein are much more energy-efficient than the wild sources, and that current technological change is toward the domesticated and hence energy-efficient direction.

But to interpret this change as one from a less efficient to a more efficient technological system would be to miss the point that the efficiency of each particular technique for protein acquisition has itself changed through time, in response to external changes which at first appear very distantly connected to protein. The localized increases in settlement density which have increased the availability of steel axes while decreasing the abundance of forest resources are the result of political and economic changes originating far from the tropical forest homeland of the Machiguenga.

The higher efficiency of domesticated protein sources is a recent development which has had other costs. Dietary diversity is probably declining; the quality of raw materials certainly is. Strange as it seems from our urbanized perspective, the Machiguenga technology is experiencing the stresses of overcrowding. Although the costs are difficult to compute with any accuracy, they are real in the subjective experience of the people and help to account for the slow, cautious pace at which they have moved to adopt what in simple energy terms seems to be a much more efficient technology.

LITERATURE CITED

Bodenheimer, Friedrich Simon
1951 Insects as Human Food. The Hague: W. Junk.

Boserup, Ester
1965 The Conditions of Agricultural Growth. Chicago: Aldine.

Childe, V. Gordon
1951 Man Makes Himself. New York: New American Library.

FAO
1965 Poultry Feeding in Tropical and Subtropical Countries. Agricultural Development Paper No. 82. Rome: Food and Agriculture Organization of the United Nations.
1964 Protein. Rome: Food and Agricultural Organization of the United Nations.

INCAP, ICNND
1961 Food Composition Tables for Use in Latin America. Bethesda: Interdepartmental Committee on Nutrition for National Defense.

Harris, Marvin
1975 Culture, People, Nature. New York: Crowell.

Johnson, Allen
1975 Time Allocation in a Machiguenga Community. Ethnology 14:301–310.

Latham, M.
1965 Human Nutrition in Tropical Africa. Rome: FAO.

Montgomery, Edward and Allen Johnson
1975 Machiguenga Energy Expenditure. Ms.: Washington University and UCLA.

Platt, B. S.
1962 Tables of Representative Values of Foods Commonly Used in Tropical Countries. London: HMSO.

Sahlins, Marshall
1972 Stone Age Economics. Chicago: Aldine.

White, Leslie
1959 The Evolution of Culture. New York: McGraw-Hill.

*

8

The Development of the Potter's Wheel: An Analytical and Synthesizing Study

ROBERT H. JOHNSTON*

The Rochester Institute of Technology, Rochester, New York

And the word that came to Jeremiah from the Lord: "Arise and go down to the potter's house, and there I will let you hear my words." So I went down to the potter's house, and there he was working at his wheel. And the vessel he was making of clay was spoiled in the potter's hand, and he reworked it into another vessel, as it seemed good to the potter to do. (Jeremiah 18: 1–4)

In 1890 Sir Flinders Petrie excavated Tell el Hesy in Palestine after ten years of digging in Egypt accumulating knowledge and experience in excavation and data analysis. His recognition of the chronological value of

*Robert H. Johnston is Dean of the College of Fine and Applied Arts and Director of the School for American Craftsmen at the Rochester Institute of Technology, Rochester, New York. He wishes to acknowledge the assistance of Ms. Rose Marie Deorr in preparing and proofreading the following article.

pot sherds in stratigraphical excavation led him to establish a "Corpus" of dated sherds. Since that time the study of ceramics has been an integral part of every excavation. He wrote in his publication *Tell el Hesy* in 1891 the following:

> Once settle the pottery of a country and the key is in our hands for all future explorations. A single glance at a mound of ruins will show to anyone who knows the styles of pottery as weeks of work may reveal to a beginner. (Petrie 1891:40)

In 1937, the late G. Ernest Wright in his thesis entitled *The Pottery of Palestine from the Earliest Times to the End of the Early Bronze Age* recognizes a much greater value in the study of pottery than typology and chronology. He stated:

> Its (pottery) greatest value at present is undoubtedly chronological; yet more exact studies in the future will perhaps allow the student of ethnology, commerce and related subjects, to make far-reaching deductions from ceramic evidence, for which at present there is so little ground. (Wright 1937:1)

Having worked as a member of Dr. Wright's American Schools of Oriental Research field staff for four years and now with his former students, Dr. Stager and Dr. Schaub, I have found it rewarding to see this dream become a reality as the ceramic specialist works with the geologist, excavator, typologist, ethnographer, conservator, and others.

The next logical step in the study of ceramic material was taken by my teacher and mentor, Dr. Frederick R. Matson, who has developed an ecological approach to the study of the ceramic yield of a given site. He stated in his book, *Ceramics and Man:*

> Unless ceramic studies lead to a better understanding of the cultural context in which the objects were made and used, they form a sterile record of limited worth. (Matson 1965:202)

Working with Dr. Matson as a student for a number of years, and being a staff member of ten expeditions to the Middle East, I have had the privilege of combining the technical and analytical study of excavated ceramic material with the study and filming of village potters.

These potters, who worked in remote areas in a number of countries, made their pottery using traditional techniques that have been handed down from one generation to another with little change. My experiences may be summed up in the phrase, witnessing the "archaeological present". (Matson in verbal presentation 1970).

Potters are conservative people who resist change and pride themselves in their tradition. Various programs which have been introduced to change potters and pottery techniques have failed. The following is a statement by George Foster based on his observations:

> My recent research in Tzintzuntzan, Mexico, where two thirds of the families earn all or part of their living from pottery-making, shows us that as a class potters are measurably more conservative than farmers or fishermen, who make up the remainder of the families. . . . There is a good deal of evidence to suggest that the conservatism of Tzintzuntzan reflects a basic

Figure 1. Woven impressions on pottery base. Early potters built their vessels on potter's bats made of woven material which allowed them to turn the piece as they worked and allowed the potter to move the finished piece to the slow drying area without deforming the piece. Potters have used woven pads, pieces of tree bark, stone and broken pottery fragments on which to build vessels. (Tell el-Kheleifeh, Jordan; Req. No. 4019, PAM No. 40.656)

> conservatism in the psychological makeup of potters in other parts of the world. As a rule-of-the-thumb guide to community development work, I would suggest that new community development programs avoid pottery-making villages as initial targets. (Foster 1962:143–144)

My studies have related contemporary craftsmen's practices to antiquity with some interesting results. One of these was the study of the development of the potter's wheel from hand-building on woven pads (see Fig. 1) to the tournette or slow wheel and then to the potter's kick wheel so well described in the Old Testament in Jeremiah 18:3. The purpose of this paper is to trace a possible development of the potter's wheel through study of ancient ceramic material and through observations recorded in remote villages of potters at work.

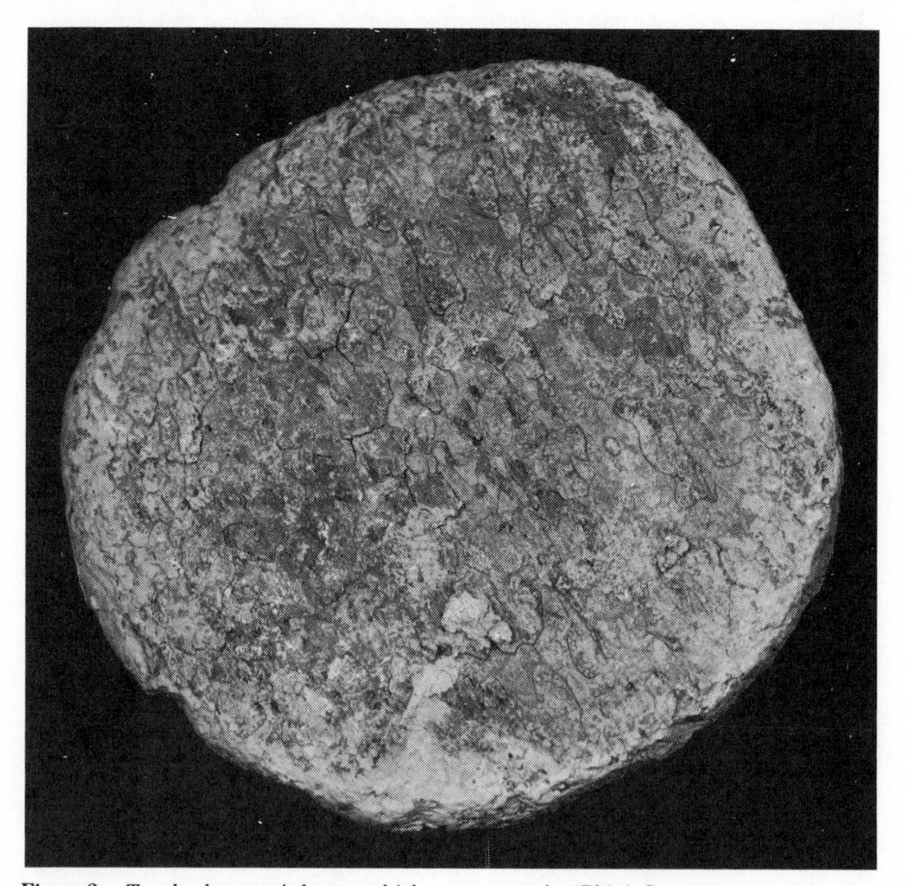

Figure 2. Tree bark potter's bat on which pots are made. (Phini, Cyprus)

The earliest materials that man learned to shape and use were wood, stone, bone, and clay. But when the first potter produced a worthwhile substitute in clay for previous containers made of wood, gourds, shell, hollowed stone, skulls, or animal skins and bladders, a technology was begun that is with us to this very day. The potter's tradition has an unbroken history from its inception to the present. While some of the potter's tools and equipment have changed, and are powered by electricity instead of footpower, the pottery industry is today almost the same as it was in antiquity.

No one can point with any authority to how the potter's art began. The myths associated with these beginnings, however, are fascinating. While time will not permit a lengthy discourse on the subject, I should like to present one of these origin statements photographed from a copy of the Sacred Book of Potters in the village of Charikar, Afghanistan:

Figure 3. Potter hand-building coil pot on curved (convex) plates.

Figure 4. Detail of convex plates allowing the potter to slowly turn the pot as she works.

> O' potter, if one comes to you and asks you a question on any occasion, that Who originated this profession. Instantaneously answer that at the time of Adam. With the order of God, Gibreil brought all the tools to the God's chosen one (Adam), listen to me O' student.
>
> Q
>
> If one again asks a question and demands an answer without delay, who perpetuated this profession after Adam?
>
> A
>
> Tell him that Noah successfully handled this craft. After him, Ibrahim knowledgeably did the job, After him, Ishaq became very fond of the profession, Then with the command of the Allah, this work was done by Moses, Then reached this profession to our Muhammad, After him, Ali, the saint of Allah, kept busy with the job. al-Hassan learned from him and made it clear to the world and became famous among all people.
>
> Q
>
> If again one comes and inquires from you. Who was the leading pioneer in the craft? Tell him that the first leading figure who was perfect in this work was Gibreil. Finally; Mir Kulal learned this profession in its entirety.

(Note: The author photographed the Sacred Book but the above translation was provided by Dr. Frederick R. Matson from his own studies in Afghanistan on July 8, 1972. The actual translation was made by Abdul Raziq Palwal in Kandahar in November 1972.) In Afghanistan some potters who claim to be descendants from Mir Kulal still award the master's belt to the potter who has finished his apprenticeship and is now a finished master potter.

The myths about who invented the potter's wheel are endless and related to: Emperor Huang-Ti in the middle of the third millenium, the Greek Talos of Crete, Hyperbios of Corinth, Koreibos of Athens. Plinius (24–79) mentions Anacharsis, a Scythian of princely rank, as the inventor of the wheel. The Sacred Book of Potters from Afghanistan states the following concerning the discovery of the potter's wheel:

> If again one is questioned and wants the account from you, as to who brought the potter's wheel for this profession?
>
> A
>
> In your reply tell him and make him feel happy. First of all it was brought from heaven by the grandfather Adam.

Figure 5. Cypriot potter hand-building large storage vessel on non-turning bat.

The potter of today gathers his clay, makes and fires his ware in much the same manner as the potter of the Bronze Age in the Middle East or the early potter in Africa or China. Potters of all times have utilized clay for its two distinctive properties: being plastic when wet and hardening when subjected to heat. Clay is readily available to potters everywhere and can be utilized to one degree or another with little if any refinement. Large stones must be removed to prevent cracks from forming as the clay shrinks around the stones in the drying process and at times nonplastic material must be added to temper the clay such as cat-o-nine-tail fuzz, dung, ash, ground up shell, and so on. Most prewheel potters work under some shade to retard the too-rapid drying of the pieces they are building. Most of us are quite familiar with the usual hand-building processes: the pinch-pot technique and the coil technique, so no time will be spent on them.

Large vessels can be made on a potter's bat (see Fig. 2, 5, and 6) by using thick coils of clay which can be worked to form the vessel as the potter walks round and round the pot. In this technique the vessel being formed remains in the same position but the potter moves. As large containers are made, hoops of wood or bamboo are placed around the vessel to prevent the pliable clay from moving and cracking. Cord or cloth is also used as a kind of belly-band to strengthen the piece as additional coils and the weight of wet clay are added. The piece is thinned and shaped by using the hands as paddle and anvil and by stroking the inside of the vessel with the side of the potter's hand while supporting the outside.

Early Palestinian potters used stone tournettes with the head and socket carefully formed in stone (see Fig. 7). Oil must have been poured in the socket so the stone head could be slowly rotated. Some stone wheels were reported still in use at the turn of this century by Miss Olga Tufnell and others who visited potter's villages on Cyprus at that time. (See other suggested readings.)

Another technique used in the Middle East during Early Bronze I, Ia and EBII was the technique of stroking clay (see Fig. 8–14). Pottery found at Bab edh-Dhra in Jordan by Albright, Lapp, and Schaub show the forming marks left by this process. The potter simply adds water to his clay, which is almost always a thixotropic clay, to make it plastic. Usually the wet clay is allowed to stand overnight to allow it to become more uniformally wet. The clay is then wedged to deaerate it by slapping it together or by kneading it. The potter makes an approximate ball of clay, places it on a bark or stone bat, and opens the clay ball by forcing her fist down into the clay. I use the feminine pronoun because every potter I have studied who works in a nonwheel technique, has been a woman.

Figure 6. Bamboo hoops placed around large storage vessel firm up piece so work can continue.

Men take over when the mechanical kick wheel or stick-turned wheel appear. As the potter opens the basic form with her fist the form is slowly turned while being paddled with the other hand to thin and keep the clay walls of consistent thickness. Potters have turned or rotated the clay on woven pads, curved bowls (see Fig. 3 and 4), tree bark bats, and a variety of styles of tournettes. After the form has been opened the potter uses a round piece of wood or bamboo and begins to stroke the clay inward and upward while supporting the inside of the thinning wall with her left hand. As the vessel grows and the neck becomes smaller and smaller the potter uses a stick to support the form. For larger taller pieces made by this stroking technique the upper necked part is made first, then the lower more rounded section and, when slightly leather hard, the two sections are joined together. Sometimes a decorative band of clay is

Figure 7. Potter's wheel; Basalt, Late Canaanite Period; Hazor Excavations. (Katz 1968:68, illustration 54)

Figure 8. Stroking technique for opening the clay form. (Phini, Cyprus)

Figure 9. Stroking technique with stick against hand. (Phini, Cyprus)

Figure 10. Stroking technique. (Phini, Cyprus)

Figure 11. Stroking technique raising with a stick. (Phini, Cyprus)

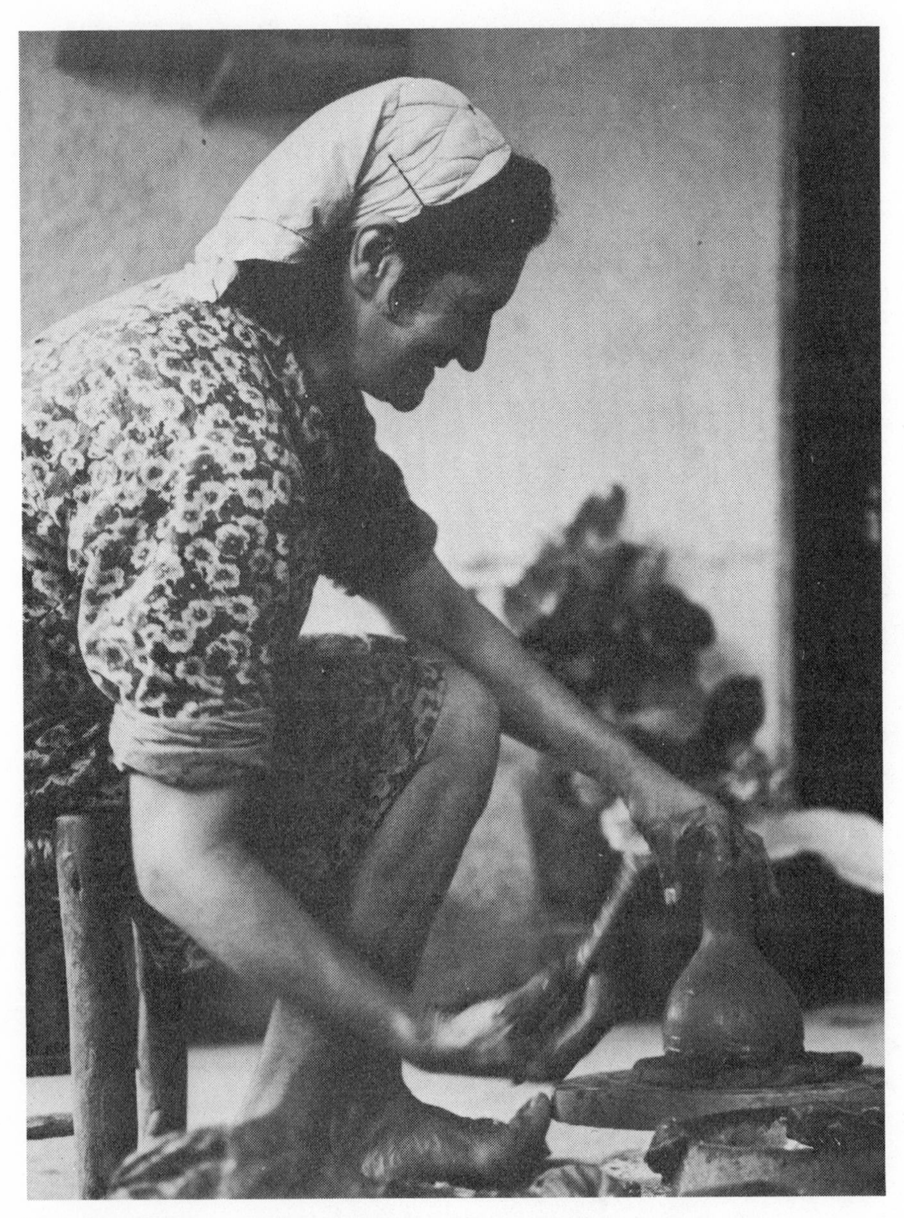

Figure 12. Stroking technique. (Phini, Cyprus)

Figure 13. Completing upper half of pot using the stroking technique. (Phini, Cyprus)

Figure 14. Completed pot formed by the stroking technique. (Phini, Cyprus)

placed around the join of the two pieces. Finished pieces of pottery made by this stroking technique were trimmed down around the base during the leather hard stage. This trimming produced a rounded bottom which worked well for vessels to be used on earthen tables, where depressions in the surface would hold the pot, or on pottery which would rest in a ring of clay or woven material of some type. At times during the finishing process, clay or wooden paddles and anvils were used to thin the bottom section and, occasionally, to apply a decoration to the piece.

The early tournette was probably quite simple: a wooden wheel turning on a stick placed in the ground or running through a supporting board fastened to the ground with pegs (see Fig. 15 and 16). Building a piece of pottery on a potter's bat placed on a platform that could be rotated slowly with the hands or feet as the potter worked was a decided advantage to the potter. While working in the Troodos Mountains (Cyprus) in 1971 and 1973 in the villages of Phini and Ayios Dhimetrios, I found one potter using a wooden wheel head that was fastened to a plank by a metal pin on which it turned (see Fig. 17A and B). I noticed a white powder being used for a lubricant and was excited to note that the powder was bread flour (see Fig. 18). What an interesting connection between clay and flour, both associated with Neolithic times. It was also most interesting to note that the village potters in these villages were making nonfunctional vessels which had designs that used the bird, tree, and snake motifs. We had studied these symbols on Early Cypriot pottery excavated from a number of sites on the island. Indications were that these techniques and symbols, handed down through many generations, were still going on.

Tournettes, of the type we are discussing, are shown in the tomb paintings and texts of ancient Egypt and are portrayed in wood and clay models (see Fig. 21).

Another tournette technique studied in detail is the use of a tournette which has a square head and which is connected to its base by a pin with grease and another with a ball bearing (see Fig. 19A & B and 20A & B). This type of tournette allows the potter to use a variety of methods to form his vessels. He can coil build and use the tournette as a simple turntable and he can true up his vessel by spinning the tournette and using his hands or a wooden spline (shaping stick) to make the outer surface more true and smooth. One potter observed using this type of tournette actually spun the piece on the tournette and, using its centrifugal force, actually formed the neck of the vessel and its rim as a potter would do using the kick wheel.

The usual tournette method can be described as follows based on a number of studies made on Cyprus. The potter fastens a bark bat or stone

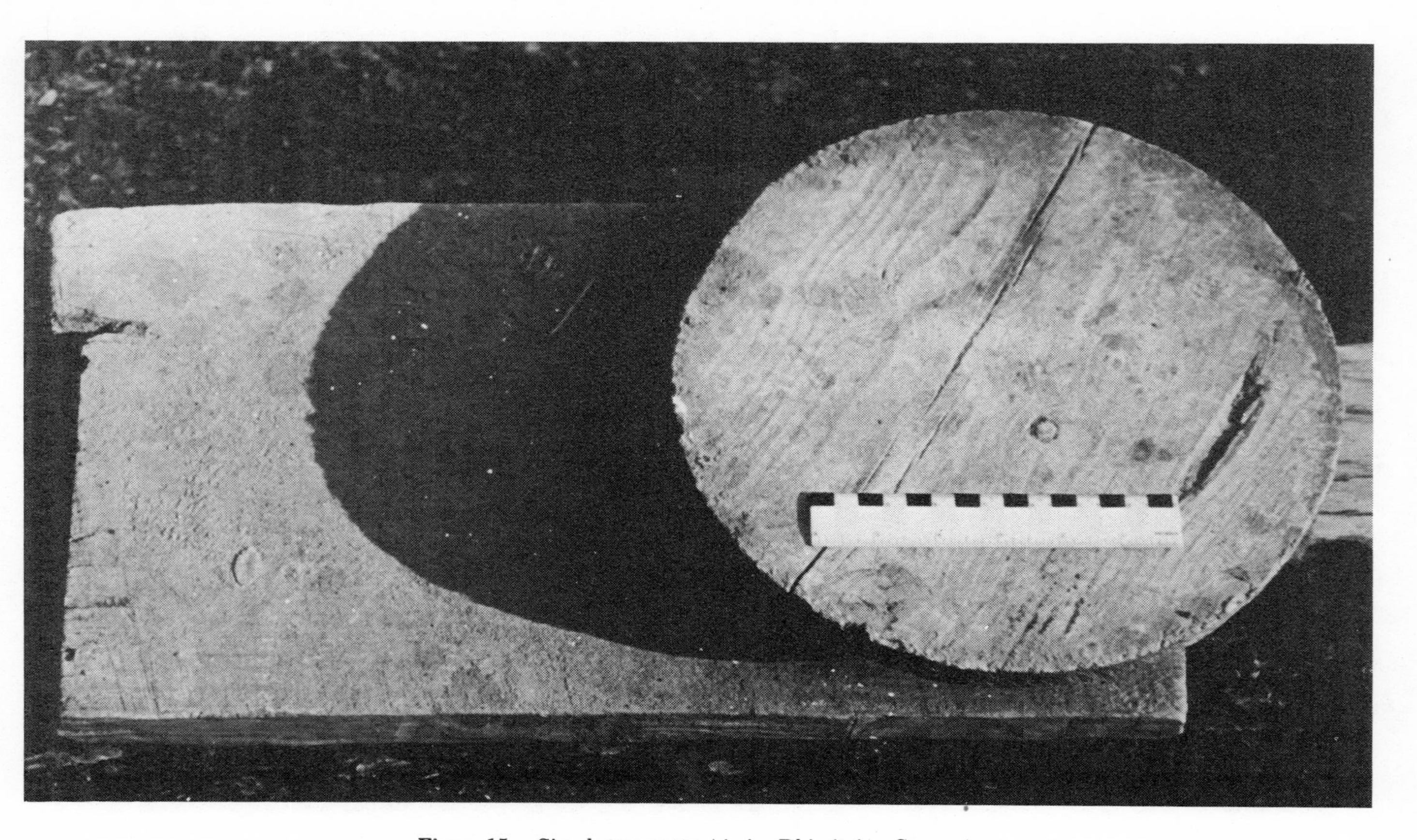

Figure 15. Simple tournette. (Ayios Dhimitrios, Cyprus)

bat on top of her wheel head upon which she'll build the vessel. She takes a lump of clay from nearby and deaerates it, usually by separating it into pieces and slapping them together. This also distributes the moisture evenly in the clay so the potter has a more homogenous material to work with. (As clay stands for any length of time, the outer areas lose moisture more quickly than the inner mass.) The potter next fixes the pug of clay

Figure 16. Tournette head connected to base with a wooden pin.

Figure 17A & B. Wooden tournette connected with a steel pin. (Phini, Cyprus)

onto the wheel and opens the mass by inserting her fist while patting the outside as the wheel is slowly turned with the foot. A large coil of clay is rolled between the hands and applied to the opened form by working it into the lower clay form. After two or more coils are added, the side of the hand is used as a kind of scraper to smooth, form, and further knit the coils. The palm of the other hand is, at the same time, patting the exterior, as in a paddle and anvil method, to refine the exterior shape and make the walls of the pot of even thickness. More coils are added and the process continued. Periodically the vessel is spun around by turning the tournette rapidly and the exterior smoothed and refining by using a stick or half piece of bamboo. As the height of the vessel grows, cords or cloth strips are wound around it to prevent sagging as the weight of new coils is added (see Fig. 22). These cords are placed around the belly of the pot before the upper part is brought in and the neck formed. They are left in place until the vessel is leather hard at which point they are removed and the lower half of the piece refined. The potter brings the sides of the piece

Figure 18. Tournette using bread flour as a lubricant. (Ayios Dhimitrios, Cyprus)

Figure 19A & B. Square wooden tournette with metal pin and lubricated with oil. (Phini, Cyprus)

Figure 20A & B. Wooden tournette with a modern ball bearing. (Kornos, Cyprus)

in or out by building his coils in or out as they are applied. Thixotropic clay is necessary for this technique because of its unique properties. As this type of clay is worked it gets more and more plastic but when it is no longer worked it sets up quickly and becomes quite firm and strong. This allows the tournette potter to keep on working rather than having to stop and wait for the clay in the lower half to set up or having to work on a number of pieces at a time as the potters on Crete do. When the piece has been completed, the potter sometimes decorates it near the neck with several bands of decoration. Our study found that in some villages these decorations were piece-mark signs that related the pottery to the potter producing them so that monies earned would be properly distributed. We also noted that potters placed two small breasts at the base of the neck of the larger vessels. We were told that these breasts were only placed on the large mature pieces. One should note that the same breast forms appear on Bronze Age pottery. After the piece has been completed, it and the bat on which it was made are removed from the tournette and put under shade to slowly air dry to the leather-hard stage. Later it is removed from the bat and the lower part of the piece shaved down using an iron shave. This trimming not only refines the shape of the lower half and

Figure 21. Egyptian potters using a tournette from the 21st Century at Sakkarah. (Pritchard 1971: photo 26)

Figure 22. Maria, Kornos potter working on a ball bearing tournette at Kornos, Cyprus. The pots have cord wrapped around them to prevent slumping. The decoration is Maria's piece-mark.

bottom of the piece but thins the lower clay walls so the piece will shrink more evenly in the drying and firing processes. Next the outer surfaces are sponged down with a damp sponge and the surface rubbed all over with a small round stick. This helps compress the outer layer of clay and produce a micro layer of fine clay on the surface brought there by the rubbing. The piece is then placed in a drying shed for several days to a week to slowly dry. When enough ware is dry for a kiln load, the pieces are removed to the kiln for firing.

One of the great advantages of the tournette is its portability. The small wheels can easily be transported from one place to another by donkey, horse, camel, or truck as well as from one part of the potter's compound to another. Nomadic potters who move from village to village changing clays with which they work and the forms they produce to adapt the pottery to the needs of the local market have been studied in Afghanistan. Can you imagine the problems this could cause future archaeologists? One man making pottery at one time but producing ware of different fabric and form as he covers a vest geographic area. Dr. Matson noted another aspect of the nomadic potters in his book *Ceramics and Man* as follows:

> Since many villagers, even today, spend their summers in tent camps in the mountains with their flocks, it is quite possible to find the same type of pottery, made by the same potters, but composed of different types of clay. (Matson 1965:211)

Another portable wheel used in many Eastern countries is the stick turned wheel (see Fig. 23A–H). This wheel turns on a wooden pin with a clay bearing which has been stuck into the ground. The diameter of the wheel is around 96 centimeters (approximately 38 inches). These wheel heads are made of fired clay or wood—the wooden ones sometimes having spokes like a wagon wheel. The potter inserts a stick into a hole in the wheel top and turns the wheel 30 or 40 times to get it spinning. The wheel will turn for almost 5 minutes without stopping. As the wheel slows down, the potter can place the stick in the hole and start the wheel turning again. The potter, in this case a man, places a pug of clay in the center of the wheel. He pats the pug of clay into place, roughly centering it and making it adhere firmly to the wheel. He starts the wheel rotating in a clockwise direction by turning it with his hands. He then uses the stick to spin the wheel. He wets his hands and by squeezing in with both hands he centers the clay. He next inserts his left hand down into the clay while

Figure 23A, B, C, D, E, F, G, H. Afghan potter using a stick-turned wheel to manufacture water jugs. (Charikar, Afghanistan)

Figure 23B.

Figure 23C.

Figure 23D.

Figure 23E.

Figure 23F.

Figure 23G.

Figure 23H.

supporting the mass with his right hand. Maximum drag is placed on the clay and the spinning wheel at this stage and it is likely that the potter may have to re-spin the wheel with the stick several times. At this point the first draw is made. While the clay is being supported with the right hand on the outside, the knuckles of the left hand inside press against the outside hand and lift upward. The centrifugal force of the wheel thins the clay and the walls are thus forced upwards as the clay is stretched. After a tall cylinder shape has been thrown, the potter pushes out the belly of the pot and chokes in the neck. Great care must be taken when choking in the neck not to get the clay too wet, or the dome below the neck will slump and fall inside destroying the piece. The potter shapes the rim with his fingers and smoothes it with a wet piece of cloth or leather. Decorative elements are applied and the lower part cut in. A string, which cuts the piece free is pulled through the clay right below the bottom and the potter lifts off the thrown form. The clay hump left on the wheel can be recentered and shaped into another piece of pottery. If a number of pieces is made from the same pug of clay, by cutting off each piece and pulling up more clay for the next one, this technique is called throwing off the hump. A skilled potter can throw several bowls, one bottle, and several cups off one pug of clay. In some villages studied in Northern Afghanistan the potter allows the finished pieces to air dry then uses a carved clay or wooden paddle and anvil to finish, shape, and thin the bottom, as well as to add an overall decoration to the pottery. Potters using this technique can make over 50 to 60 large storage vessels in a morning. To make their clay work better on the stick-turned wheel and to control drying shrinkage, cat-o-nine-tail fuzz is wedged into the plastic clay. The stick-turned wheel appears in Afghanistan, India, Africa, and other eastern countries (see Fig. 24).

The potter's kick wheel is the wheel now most used by potters in the western world and by many production potters in the East (see Fig. 25). The kick wheel is a double wheel on a shaft mounted on a floor-bearing of some kind and supported in a vertical manner by a frame of clay, wood, or metal. The potter kicks the wheel and keeps it turning with his right foot and while sitting down keeps the wheel spinning. Usually the lower kick wheel is very heavy (at times weighing 150 lbs.) and acts as a flywheel. However other kick wheels have light wooden wheels which the potter must constantly turn with his foot as he works. The main advantage of the kick wheel is the ease with which the potter can spin his wheel and the uninterrupted time he can keep on producing pottery. The process is almost the same as that of the stick turned wheel. Some potters throw vessels upside down making and closing the bottom first then using a chuck (see Fig. 26) in which to fit the completed base, and throwing the upper part of the vessel the next day. This practice is common in Jordan and in Egypt. Sometimes the necks are added on later as separate pieces by adding coils to the thrown forms and then raising the neck by the usual throwing process. The kick wheel is a more permanent potter's wheel which cannot easily by moved from one place to another and usually is part of a potter's workshop. Again, men use this wheel since it is a mechanical process. Village potters who use this wheel use very few tools other than their hands. They use a wooden spline or stick, a cut off string, a knife for trimming, and, at times, a roulette for decorating. The potter holds the roulette against the turning pottery on the wheel and, as the roulette wheel turns with the vessel, a decorative band is impressed into the clay.

The development of the potter's wheel took place over a long period of time. In the Middle East there is little evidence of a potter's wheel until the Middle Bronze Period (*ca.* 2250–1950 B.C.). Prior to that the results of my own studies indicate much use of the tournette or slow wheel for a very long period of time. There is a great need for more study in this area and perhaps our new expedition and work at Bab edh-Dhra and the Five Cities of the Plain in the lower Dead Sea area of Jordan will help shed some new light on these problems.

Figure 24. Stick-turned potter's wheel photographed in India. (Rieth 1939:99)

The continued work of ceramics specialists combined with ethnographic studies and the study of excavated material should help add additional knowledge about the technology of ancient man and the ecological environment in which he lived. One feels a sense of urgency when making studies of village craftsmen in the Middle East and in Afghanistan. Modern technology is taking over in the making of goods and craftsmen are not teaching their sons and daughters to follow the old traditions. While we in America are in the posttechnological phase and know too well the effects on living brought on by modern technology, most small nations in the Middle East are seeking technology to help them grow and provide much needed goods and services. Ethnographic studies of village potters will have to be made in the next 10 to 20 years. After that time, much of what we now study and record will have vanished.

Figure 25. Contemporary potter's kick wheel. (Afghanistan)

Figure 26. A potter's chuck in which the neck of a thrown pot is placed upside down, so lower part can be trimmed. (Afghanistan)

LITERATURE CITED

Bible

All quotations are from the Oxford Annotated Bible, Revised Standard Version.

Foster, George M.

1962 Traditional Cultures: and the Impact of Technological Change. New York: Harper and Row.

Katz, Karl, P. P. Kalaue and M. Broshi

1968 From the Beginning. New York: Reynal and Company Inc.

Matson, Frederick R.

1965 Ceramic Ecology: An Approach to the Study of the Early Cultures of the Near East. *In* Ceramics and Man, F. R. Matson, ed., Pp. 202–217. Chicago: Aldine.

Petrie, W. M. Flinders

1891 Tell el Hesy (Lachish). London: Committee of the Palestine Exploration Fund by A. P. Watt.

Pritchard, James B.

1971 Ancient Near East. An Anthology of texts and pictures. Princeton, N. J.: Princeton University Press.

Rieth, Adolf

1939 Die Entwicklung der Topferscheibe. Leipzig: Curt Kabitzsch, Verlag.

Wright, G. Ernest

1937 The Pottery of Palestine from the Earliest Times to the End of the Early Bronze Age. New Haven, Connecticut: American Schools of Oriental Research.

OTHER SUGGESTED READINGS

Balfet, Helene

1962 Ceramique ancienne en proche-orient Israel et Liban, VI-III millenaires, etude technique. Unpublished Ph.D. dissertation, Paris, France.

Combes, Jean-Louis

1967 Les potiers de Djerba. Tunis: Publication du centre des arts et traditions populaires.

Hampe, R. and Adam Winter
1962 Bei Topfern and Topferinnen in Kreta Messenien und Zypern. Mainz: Rudolf Habelt, Verlag.

Johnston, Robert H.
1974a The Biblical Potter. *In* The Biblical Archaeologist, 37 (4), December 1974, Pp. 86–106. Cambridge, Massachusetts: American Schools of Oriental Research.
1974b The Cypriot Potter. *In* American Expedition to Idalion, Cyprus, First Preliminary Report Seasons of 1971 and 1972. Pp. 130–139. Cambridge, Mass.: American Schools of Oriental Research.

Taylor, J. du Plat and Olga Tufnell
1930 A Pottery Industry in Cyprus. *In* Ancient Egypt, Part IV, December 1930, Pp. 119–122. New York, British School of Archaeology in Egypt: Macmillan and Company.

Ziomecki, Juliusz
1964 Die Keramischen Technikin im antiken Griechenland. Raggi, Zeitschrift für Kunst Geschichte and Archäologie 6:1–31.

9

Yalálag Weaving: Its Aesthetic, Technological and Economic Nexus*

CAROL F. JOPLING

Peabody Museum, Harvard University, Cambridge, Massachusetts

When evaluating Western or European art, we tend to judge the artist on the basis of his work, to equate the quality of the art object with the aesthetic ability and technical capacity of the artist. Moreover, it is assumed that contemporary artists are motivated and choose to practice art primarily for aesthetic reasons and because of their talent. The artist's aesthetic ability and technical capacity also affect the quality of art

*Paper prepared for presentation at the AES Symposium on Material Culture: Styles, Organization, and Dynamics of Technology, Detroit, April 3–5, 1975. The fieldwork on which this paper is based on was done in April and May 1969, August to January 1970–71, and April to July 1971. I acknowledge with gratitude the support of an Organization of American States fellowship and a University of Massachusetts Faculty Research grant during this period.

objects produced in folk/primitive societies, but judgments made on this basis alone do not account for the social and economic factors which intervene. Art objects produced in these societies are often the creations of many hands, the results of collective efforts. Indeed, economic imperatives underpin the production of many of the arts, particularly the so-called utilitarian, and, in some instances can even override individual aesthetic motivations. This paper examines the system comprised of the interaction of the aesthetic, technological, and economic activities and processes which contribute to the production of Yalálag textiles, focusing primarily on the factors necessary to the creation of a finely woven product, in the belief that aesthetic quality results as much from the systemic nature of this art process as from individual talent and aesthetic ability. The discussion generally follows a factorial model (Edwards 1968:450), but shifts briefly to a decision model as a means of interpreting the choices a weaver makes.

BACKGROUND

Yalálag is a somewhat isolated Zapotec town, located in the Sierra Júarez, a long day's bus ride from the city of Oaxaca in southern Mexico. Its population of 2,600 is large enough to have insured the continuance of many ancient traditions like its market and fiesta cycle, and at the same time, to have attracted various federal services in addition to the school, including the regional electrification office, tax agents, and postal and telegraph service which keep it in touch with modern Mexico. Yalálag's subsistence has always been based on the *milpa* cultivation of corn, beans, squash, and chile, but because its land holdings are small in proportion to its population size, it has also depended on the craft occupations of weaving, and since colonial time, *huarache* (sandal) making.

Weaving on the backstrap loom (Fig. 1) existed in the Sierra Júarez in pre-Columbian times, and, in fact, the production of *manta,* an unbleached muslin-type textile, was, for Yalálag, a principal means of subsistence, second only to agriculture. A large proportion of the population was engaged in the buying and preparation of raw cotton, its spinning and weaving, and the export and trade of the cloth. In the 1920s, the arrival of factory-made cloth ended the *manta* industry, and shortly after, the introduction of commercial thread replaced spinning. Finally, in the early 1940s, the Singer Company brought in the sewing machine. With these changes, the sewing of men's and some of women's clothes became the favored industry and weaving was downgraded in importance. Today, only about one hundred women weave, the majority of them old, and/or poor.

Figure 1. Yalálag weaver weaving in standing position.

Considering these limitations, it is remarkable that Yalálag weaving still persists so strongly, especially when it has almost disappeared elsewhere in the Sierra Júarez. There are complex and multiple reasons for its continuance, but two factors seem particularly salient. First, because of Yalálag's relative isolation and large size, about two to three hundred women have continued to wear the ancient *huipil* (tunic-blouse). (See Fig. 2) Second, the *traje* or folk costume modification of this basic dress is widely recognized as one of Mexico's most attractive native costumes (Fig. 3). Throughout Meso-America, during and since pre-Columbian times, differing dress has served not only to identify residence and status, but also to symbolize the ideals of the village. Still subscribing to these notions, Yalálag traditional women, especially those who have not attended school, wear the ancient *huipil.* In recent years, wearing the *huipil* locally has been discouraged among school girls and derogated as

Figure 2. Yalálag women, dressed in traditional *huipiles,* shopping at the Yalálag weekly market.

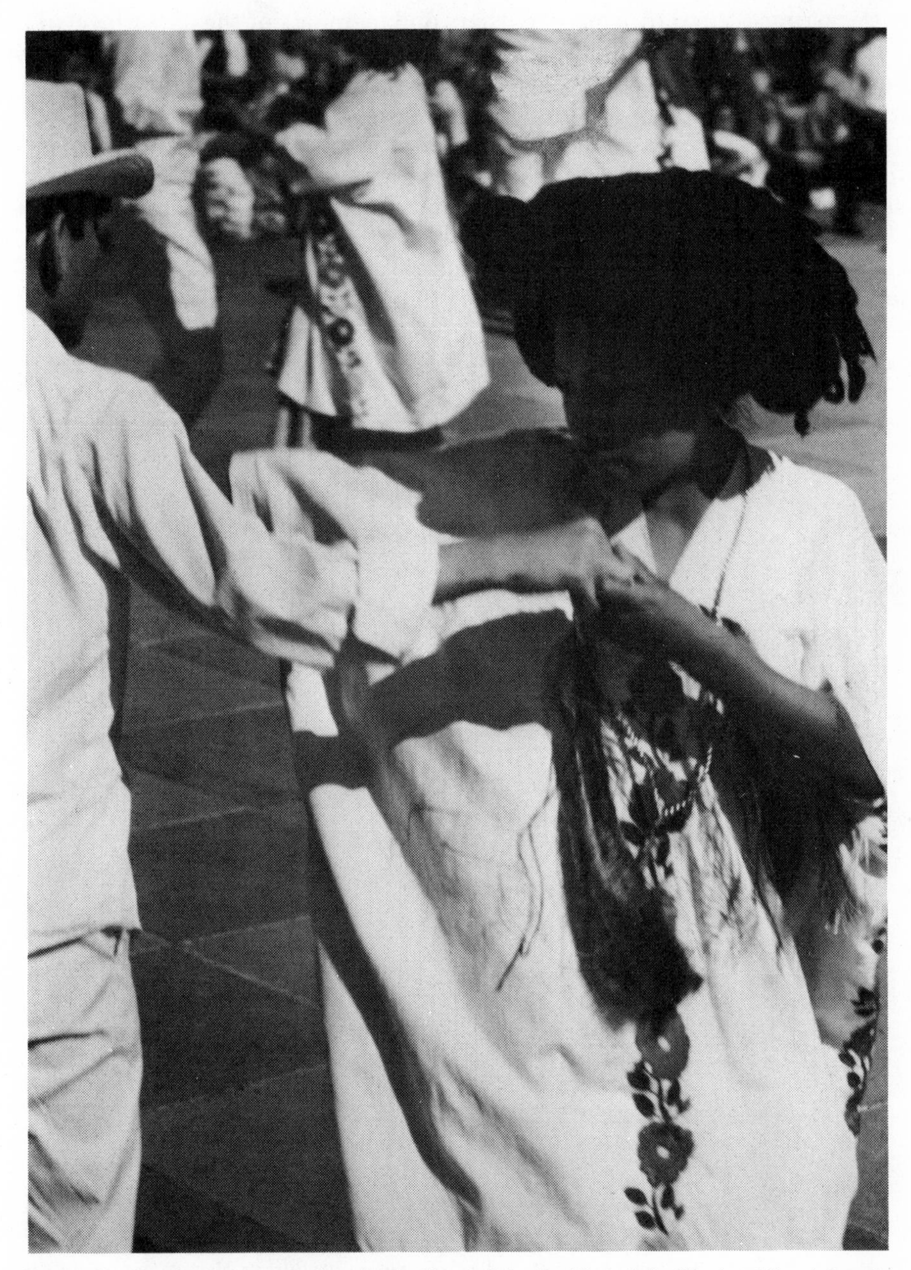

Figure 3. Jarabe dancers performing in front of the Basilica of the Virgin of Gaudalupe in Mexico City, November 1970. (Gil Mull photo)

rustic or uncultured by school teachers and members of the upper level of Yalálag society. But paradoxically, these same people understand the *huipil's* value as folk art, encourage its use at folk festivals, and, in many indirect ways, support weaving, thereby assuring its continuation. The *huipil* is thus both an ancient and modern symbol of Yalálag; one that fits both the traditionalists and the progressives. Within this ambivalent situation, some women still weave, either making garments to wear themselves or to sell as folk art. Folk art is defined here as the continuance of a traditional art form whose original utility and/or meaning has been lost or transformed; thus it is reduced to merely an aesthetic object.

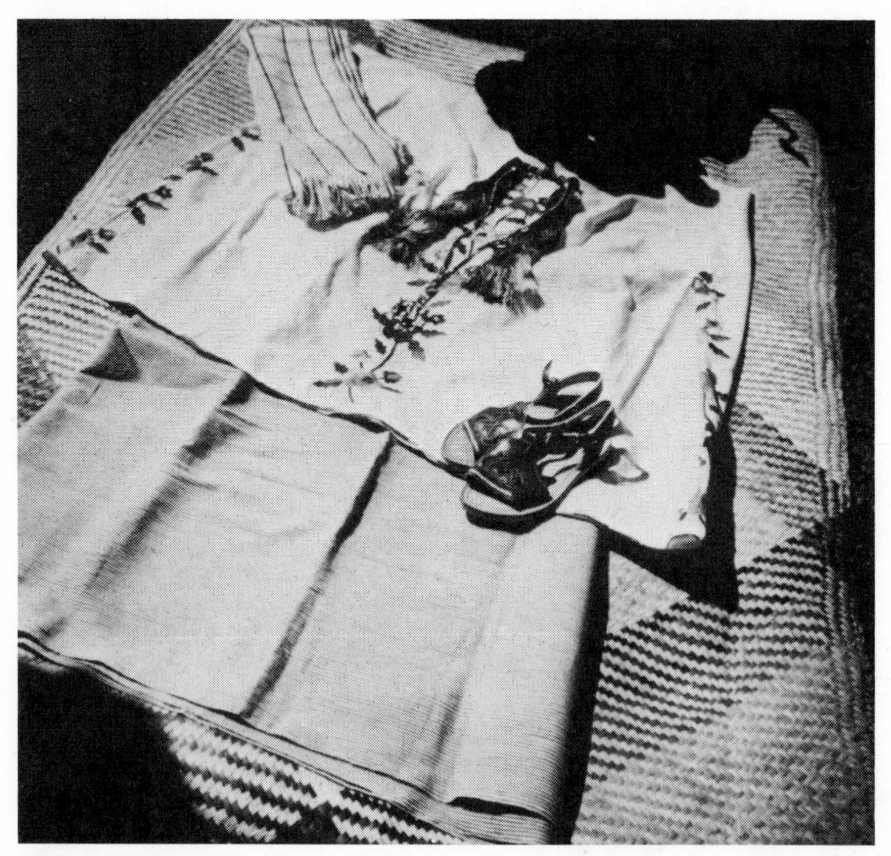

Figure 4. Yalálag *traje* or costume, including the *rodete* headdress, the *sábana* (shawl), the costume-*huipil* (tunic-blouse), the *falda* (skirt) and a pair of Yalálag designed and produced women's *huaraches* (sandals). (photo Telesforo Matias Rios)

DESCRIPTION OF TEXTILES

The woven products still made in Yalálag include the two versions of the *huipil,* a *falda,* or skirt, a *cinedor,* sash or girdle to hold the skirt in place, and three different weaves of *sábana* or shawl (called *rebozo* elsewhere in Mexico) which are frequently further modified by the insertion of colored thread into the warp and weft to make simple plaid patterns.

The floral embroidery along the side, front, and back seams which distinguishes the costume *huipil* (Fig. 4) developed when embroidery was introduced by cultural missionaries in the late 1930s. Before, only young girls had color decorations on their *huipiles,* bands of orange and red around the neck, while women's *huipiles* were pure white. Before the development of the flowered seams, a line of colored squares was embroidered onto separate strips of cloth which were then attached to the *huipil* seams. The *huipil* hem now follows this schema and is whipped with one to two-inch sections of differing colors. The folk costume, complete with its *rodete* headdress—in ancient times the most distinctive feature of Yalálag dress, but now worn only as part of the costume—is worn solely in Yalálag by some brides as a wedding dress and by local girls when they perform traditional dances either at secular occasions in Yalálag or at folk festivals in other parts of Mexico.

The traditional *huipil* (Fig. 5) is essentially the same garment, except that its embroidery consists only of a small flower at front and back just above the hem line. Both traditional and costume *huipiles* are made by the same weavers and their weaving pattern is identical, consisting of plain weave with bands of weft float patterns on the shoulders and a similar band about an inch above the hem line. *Huipil*-wearers usually possess two *huipiles,* one for daily wear and a second for fiestas or special occasions. Some women own a third, machine-made of unbleached muslin for hard household work or planting in the fields.

The remaining parts of the Yalálag dress are the same for both folk costume and daily wear. The brown and white stripes of the plain weave *falda* were once the colors of the native cotton, but now the brown cotton color is simulated by dyeing commercial thread in a solution of *encina* (oak) bark. The ample length of the *falda* is worn by lapping it in back and bunching the fullness in folds or pleats in front and holding it up with a tightly cinched *cinedor.* The *cinedor,* the piece beginners usually learn to weave first, is plain white weave. The three *sábana* weaves of gauze, *panel de abeja* (honeycomb), and *rayas* (lines) have weaving patterns created by warp manipulation. Both *sábanas* and *cinedores* have their fringes or ends worked into *punto,* or a variety of finger weave designs (Fig. 6). *Sábanas* are worn by both *huipil* and dress-wearing women as head coverings, burden and baby carriers, and as wraps, and each woman

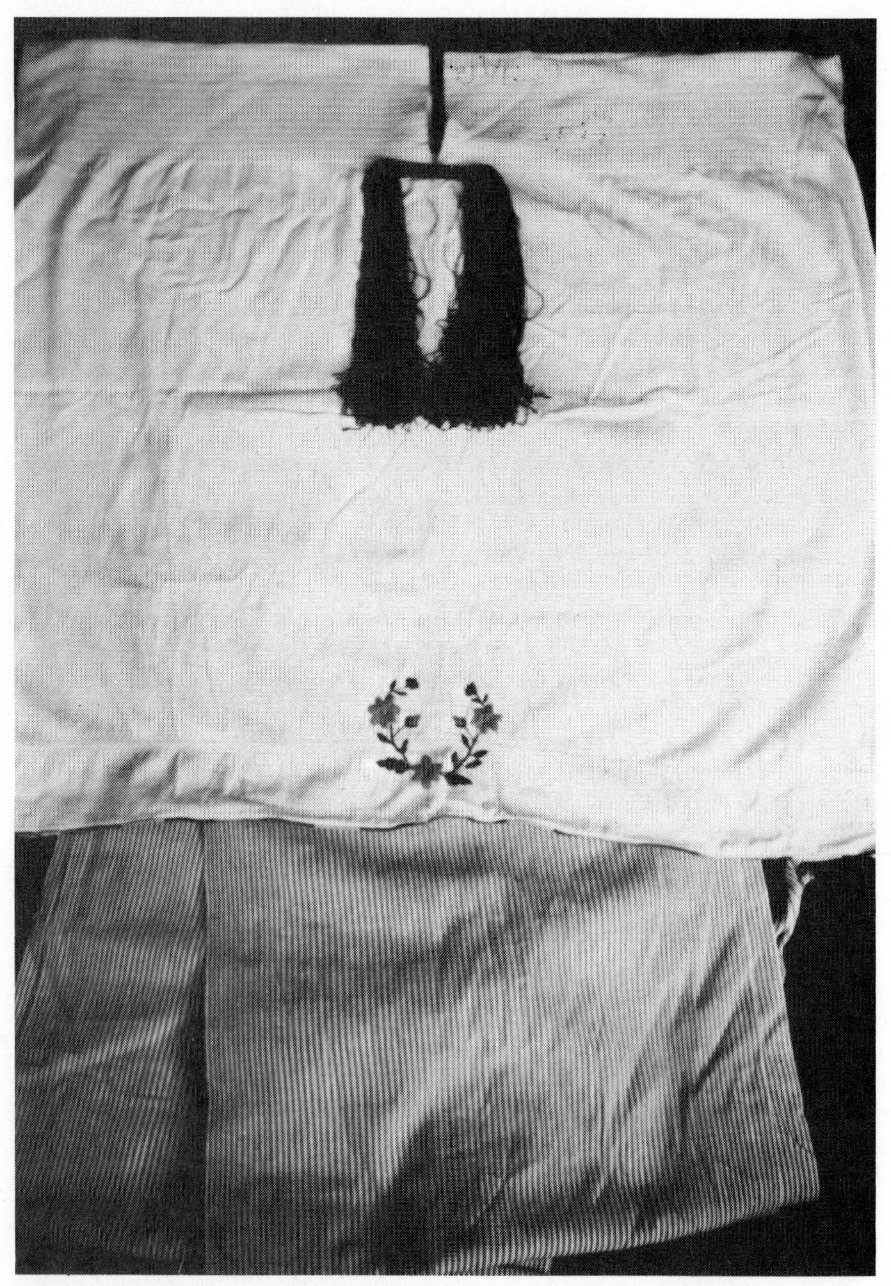

Figure 5. Traditional Yalálag *huipil* (tunic-blouse) with unusual weft float border design, and *falda* (skirt).

owns more than one for different purposes. Like the costume *huipil,* plain gauze *sábanas,* and to a lesser extent the other *sábana* weaves, have also become folk art objects and are made in quantity for export and sale to tourists in Oaxaca.

AESTHETICS

Such a limited repertoire seems to offer few opportunities for variation in aesthetic quality, but close inspection reveals a wide range of differences. Yalálag patrons and weavers account for this range in their three categories for weaving evaluation: *fino, regular,* and *corriente. Fino* (fine) requires the use of a fine thread, tight, firm, smooth, even weaving without error. *Fino* reflects Yalálag's cultural value of hard work, since tight weave requires hard and strenuous pulling on the batten. A good woman is one who works hard all the time. *Regular* (good, fair) means the use of fine thread, sometimes coarse if the weaving is otherwise very good, smooth, somewhat *delgado* (thin), insufficiently firm, but still without error. *Corriente* signifies poor, the use of coarse thread uneven

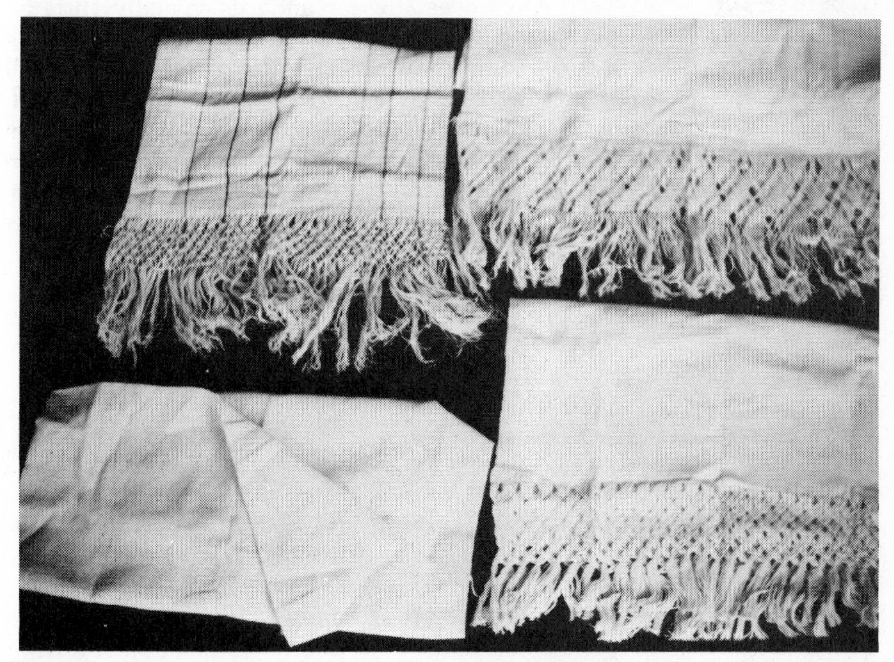

Figure 6. *Sábanas* with varying *punto* (finger weave fringe) patterns.

weaving, sometimes loose, sometimes tight, with uneven edges and many errors. Qualifying words such as *muy* (very), tone of voice, and gesture can further shade these evaluative distinctions. Weaving must meet high standards to be considered of the best quality, but taken together, these categories indicate that the expectations are for a range of aesthetic quality.

Colors are evaluated separately and their initial impact may, at first, override other impressions. Strong-hued combinations of the primary colors: red, yellow, blue, and, in addition, pink, are preferred, with bright green and turquoise also selected frequently. These strong, bright colors are said to reflect happiness and good health. Pale colors suggest sickness and weakness and are not admired by Yalálag people. White is the color worn by all Zapotecs in the Sierra Juárez, and *puro blanco* (pure white) has a significance which transcends the notions of purity, goodness, and cleanliness. It is the ideal color for Yalálag people and at its most gleaming seems to have for them a lustre of its own, one from which a special quality emanates.

These aesthetic rules constitute the standards for Yalálag weavers. As in traditional societies elsewhere, the standards themselves are basic cultural elements which are transmitted through the learning process (Ong 1969:638–40). A weaver experiences these standards visually through daily contact. The weaver's capacity to recognize the distinctions among woven objects, that is, for example, to perceive the perfection of detail which is essential to fine weaving is intrinsic to her aesthetic ability. Differential aesthetic ability accounts in some measure for variation in weaving quality, but the degree of adherence to the rules is also influenced by the weaving technology and the constraints of economic production and distribution, as well as a multitude of other social factors beyond the scope of this paper (Jopling 1973).

TECHNOLOGY

The technology of weaving is comprised of three elements: three simple instruments, the loom, reel, and *tabla* (warping board); some materials, the thread; and some specialized knowledge, the patterns and procedures of weaving (Merrill 1968:576). The loom (Fig. 7), essentially a collection of sticks of varying sizes and weights, is devised to hold thread so it can be manipulated into cloth. It is basically an extension of the weaver's body, so that she functions both as actor and instrument during the weaving process. Although the procedures of backstrap weaving are

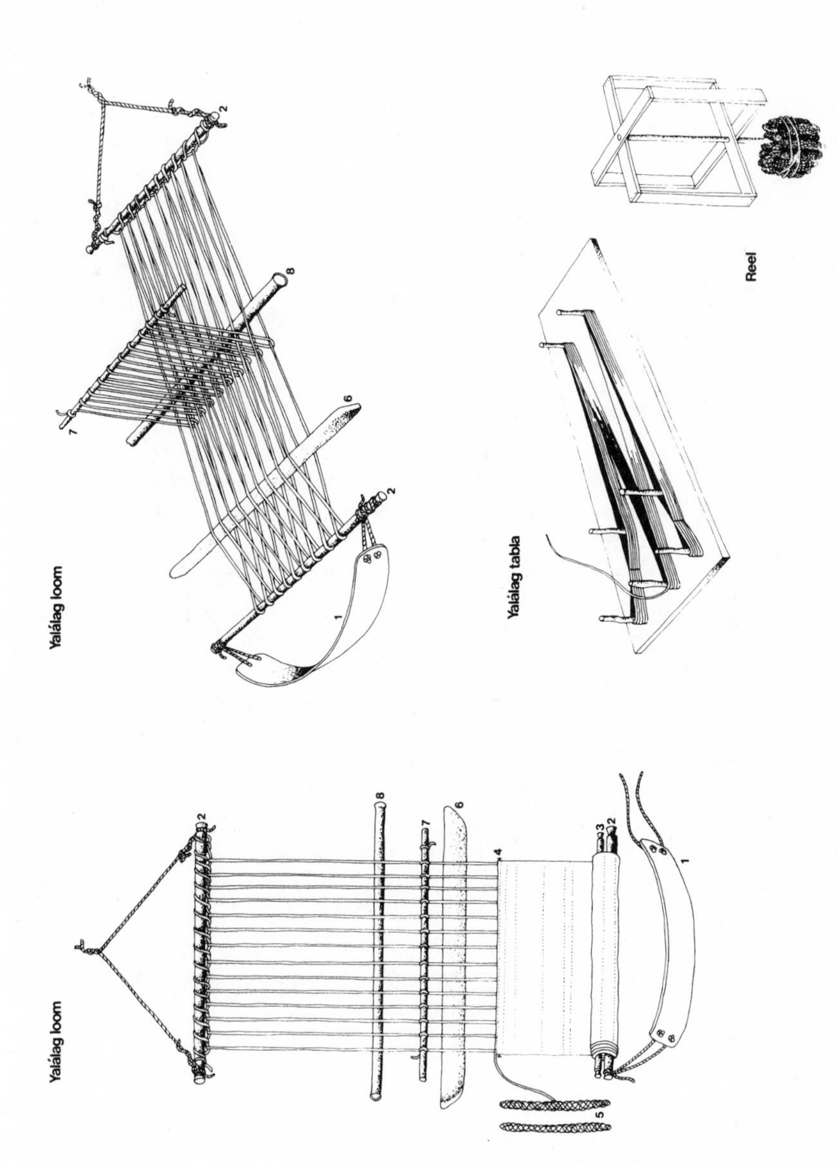

Figure 7. Yalálag loom, *table* (warping board) and reel. (Drawings by Jeanne Taylor)

well known, they are included again here to emphasize both the multiplicity and complexity of the body motions and the degree of time required for even plain weave.

Procedures

The procedures of weaving occur in stages, which are useful for scheduling time in conjunction with other household and family needs. Puting the warp on the loom and inserting the heddle are best done without interruption, but otherwise, a weaver can stop her work at will and not damage the weaving quality.

Warping. Setting up the loom begins with the preparation of the warp. Cotton thread, purchased in skeins which are sold by the *libra* (pound) is wound in two bands on the reel. Using a thread from each band to make the warp thread double, the thread is wound from the reel onto the warping board in a kind of figure eight pattern, making a cross between the upper and center outside pegs. When the warping is completed, the weaver ties a string through the warp threads at the cross to keep them in position, folds the warp over itself, and to give it body, puts it into an *olla* (jar) filled with water and *masa* (corn meal paste) to soak overnight. The following day, the warp is taken from the *olla,* arranged on the two end bars and left in the sun to dry for half a day.

Setting up the loom. To set up the loom for weaving, the warp threads are first put in order on the upper end bar. A cord is then run through and looped in a spiral through and around groups of warp threads, dividing them by eye, not by count, into about twenty to twenty-four sections. Afterwards, the loom is hung on the wall in weaving position, and the weaver sits on the floor or on a low stool with the lower end bar in her lap and begins to straighten each warp thread, lifting the thread and adjusting it onto the lower end bar to make sure it is not crossed with another warp thread and that the threads are kept clearly separated into upper and lower warps (the space between is still secured by the loom string). When the threads have been adjusted, the batten is placed in the shed (the space between the warps).

Next, the heddle is inserted by holding it in the left hand and with the right hand looping a continuous length of cotton sewing thread around the heddle stick, then making a twist before looping it again around each thread of the even-number warps below (Fig. 8).

To prepare the shuttle, the weaver winds two threads together, one from each band on the reel, onto a thin, light stick, twisting two or three turns at one end, drawing the thread down to the other end in a spiral, giving two more twists at the other end, and spiralling back up so that the shuttle thread is evenly distributed. The weaver accomplishes this task very rapidly with a motion that looks like a simple figure eight pattern.

Weaving process. When the weaver begins to weave, she attaches the loom to the wall or tree, adjusts and ties the backstrap, braces herself with her feet, and leans against the backstrap to keep the warp taut. She begins to weave from the left; raising the even-number warps by drawing up the heddle stick, she inserts the batten on its edge in between the two warps, passes the shuttle through the shed, leaving one pick of weft, turns the sword on its flat side, brings it with its thin edge toward her to

Figure 8. Weaver inserting heddle. The reel and shuttle are behind her to the left; the *tabla* (warping board) leans against the wall to the right of them.

drive the weft-pick down to the working edge, removes the sword from the shed, draws forward the shed roll which opens the counter shed, with the odd-number warps forming the upper layer and inserts the sword on its edge to keep the space open, passes the shuttle through from right to left, leaving pick behind, turns the sword on its flat side to batten down this pick, removes the sword from the shed. These procedures are for plain weave and are repeated until the cloth is finished.

The weft float patterns characteristic of the *huipil* shoulder and the *panel de abeja, rayas,* and gauze *sábanas* require additional manipulation of the warps with the use of a second heddle. Color is added by interspersing bands of colored thread into the warp at intervals which are estimated by eye or measured by four fingers or a hand's width. Several small shuttles wound with colored thread are used to add color to the weft. Three picks of color are inserted for each band of weft-color and the spaces between are organized approximately the same way as the warp-color.

Other Technological Procedures

Punto. All *sábanas* and most *cinedores* have their ends or fringe threads finished with *punto* or finger woven patterns, done by women specialists, usually not the weavers themselves. Some women know all the eight named designs, others only one or two. To make *punto* a woman sits on the ground or on a low stool or chair to twist and loop the threads into intricate patterns while the cloth is held on a table with a stone or flat iron (Fig. 9).

Embroidery. Plain satin stitch is used to fill in store-bought stencilled patterns transferred to a *huipil* with carbon paper. A few embroiderers devise their own designs (Fig. 10).

Sewing. *Faldas,* which are woven in two lengths, are flat stitched by sewing machine along their lengths. Both *faldas* and *cinedores* (if without *punto*) are machine hemmed at the ends. *Huipiles,* woven in one length, are cut in the center, and then the seams are joined without overlap with small invisible hand stitches to make the weaving look continuous. The hems of *huipiles* are whip-stitched with short blocks of varying colored thread. Nonweaving specialists perform these tasks.

Dyeing. One woman dyes all the brown cotton *falda* thread. She soaks the thread for a week in a solution of *encina* (oak) bark, dries it in the sun, resoaks it for two more days, and then dries it again, this time in the shade.

Bleaching. After finishing, but before embroidery, garments are soaked in a commercial bleach solution for two or three days and then sun dried.

Tassels. Store-bought skeins of Japanese silk are cut into eighteen inch lengths and braided into plain or varicolored tassels to be sewn to the front and back neckline of the *huipil.* Nonweaver specialists make these, or they can be purchased ready-made at the store.

Knowledge

The foregoing technology is the same for all weavers. At the gross level: the motions, the stance, the way the loom is set up, the procedures for weaving do not vary. Differences in weaving technique occur only in fine adjustments, the skillful matching of threads, the measurements, the attention to such details as keeping the edges even and the pattern regular. The learning process is the source of this conformity. In traditional societies, the transmission of technology like the transmission of

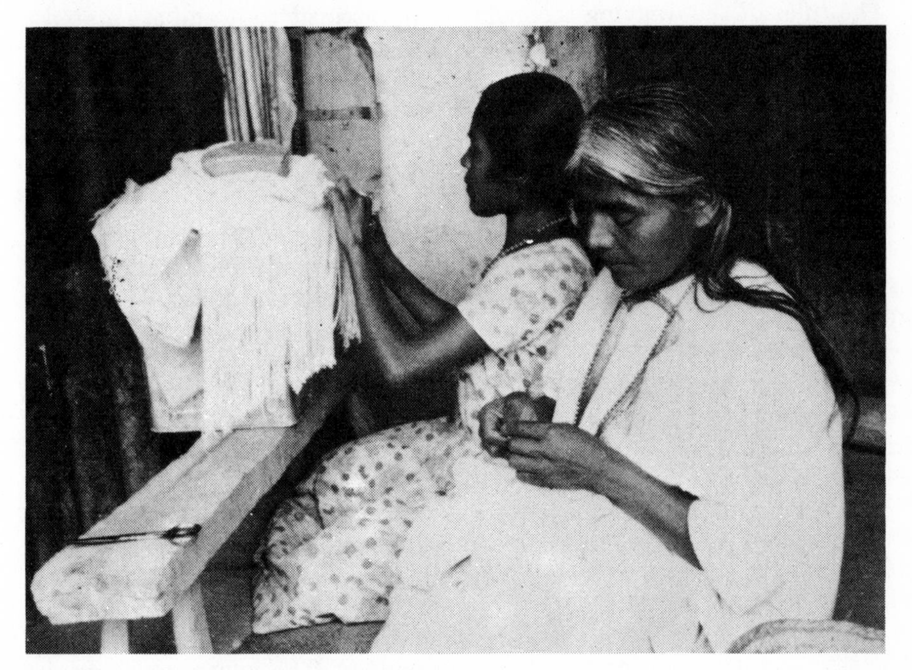

Figure 9. The woman in the foreground is sewing a *huipil;* the woman behind is making *punto* (finger weave fringe).

other aspects of culture is accomplished orally and through observation. There are no written instructions or recipes which permit differential interpretation of method. Like folk tale, visual art and its methodology is learned by imitation and repetition. To simplify transmission, themes are limited and conformity is valued over deviation (see also Ong: 1969). A young girl watches her mother at her weaving until she believes she can weave, that she has mastered the art. Confident that she can proceed without faltering, she begins to weave, and only then is she given verbal instruction in the fine points, setting-up procedures and other specific information. The novice weaver's ability to perceive and imitate the motions of the accomplished weaver and to internalize them, to coordinate her perceptions with action, account for the degree of her technical competence. Weaving is thus learned by observation and action as a motor activity, at the haptic level, more like a sport such as golf or tennis in contemporary society than as an intellectual exercise; and the rhythmic motions become integrated into the unconscious of the weaver. A good weaver, like a good tennis player has the necessary eye-hand/body coordination, has "a feel" for her craft, and is not "just going through the motions."

The rules of the weaving technology are intrinsic to it and are stated through the learning process. There is only one way to weave, and there are only certain objects and patterns to be woven. The specialization of the technology system also reinforces conformity. The most competent specialists—embroiderers, *punto*-makers—are known and their work is sought after or emulated. Deviation is not admired except as it is manifested in perfection.

The rules of technology, like the aesthetic rules, set high standards for Yalálag weavers. Coordination, health, good eyesight, and strength are all requisites for competent Yalálag weaving set by the technology. The physical demands of standing, of extensive arm motions, of the ever-increasing weight of the loom are clearly beyond some who are old, sick, or malnourished. The relatively simple patterns and the rhythmic motions of weaving make it seem automatic to the uninitiated, but even minimal inattention can cause broken threads and loops in the weft and the more complex patterns require counting and even mathematical calculations.

The aesthetics and technology of weaving are interdependent. Some of the technology has developed in response to aesthetic imperatives,—garments are bleached to achieve the desired *puro blanco* whiteness and to blend in errors (Leene 1967); a firmer cloth is attained by weaving in a standing position—but, at the same time, the aesthetic category of *fino* reflects the optimal capability of the technology. The colors, which were introduced in the 1930s, apparently have not become fully integrated into the weaving aesthetic. They are often arranged asymmetrically, possibly expressing a nonweaving aesthetic, in contrast to the symmetric order of the weaving patterns, which may result from a technological imperative.

The aesthetic and technological rules are either fixed or have their parameters defined or stated. Differential conformity to them is, in part, idiosyncratic, a result of variation in individual ability. Simply stated, a weaver can only make a *fino* textile if she is able to coordinate her aesthetic ability—her perception of what constitutes fine weaving with her technical capacity—her perception of how to accomplish it. The economic system, on the other hand, although it exerts constraints on even *fino* weavers, has an inherent flexibility which permits the accommodation of weavers of varying competence through its multiple channels.

ECONOMIC PROCESS

There are three aspects to the economics of weaving: 1) the weaver's economic situation; 2) the input, or the investment in time and materials necessary for weaving and finishing a product; and 3) the output, or the products' sales and earnings. The weavers' economic situation ranges from the poorest at subsistence level or below to property owning individuals with some savings, but never to the upper level of Yalálag society. Input and output data are correlated in the Table 1 below.

Economic System

The flow chart (Fig. 10) illustrates the economic system including the roles of weavers and other individuals who participate in the creation of Yalálag textiles, and the various channels through which these products pass. To begin with, the weaver (W), first acquires by inheritance or purchase, at a cost of no more than fifty pesos, a loom, warping board, and reel (L, T/R). Then, having decided what to weave, a step to be discussed in detail below, she obtains thread, either by purchase or on credit from the *tienda* (store) (T), is given it by a patron-buyer (Pa), for whom she weaves as a member of a putting-out system. After completing the product (P), she keeps it for herself (S), or sells it (Yalálag village sales, YVS), to one of the following: 1) a patron-buyer (Pa) who has contracted for it, or who pays for the weaver's labor if thread was provided; 2) to the *tienda* (T) to fulfill a contract, or to pay off credit; 3) to an embroiderer (E); 4) to an individual (I). From this point on, except for those who have produced garments for themselves, weavers are no longer involved in the processing or sale of their textiles, although these products must still go through several hands before their completion and final sale.

TABLE 1 Cost in Time and Materials Correlated with Selling Price

Objects & Processes	Time of Producing	Material Cost in Pesos	Selling Price in Pesos
1. *Huipil*			80–100
weaving	5–7 days	1-1/2L. = 12	40–60
bleaching	1 day	bleach .5	2
sewing (whipstitch)	1–2 days	thread 1.5	6.5
embroidery	1 day	thread 1	6
tassels	1 hour	silk 7	7.5
Costume-huipil			120–150
embroidery	5–8 days	28.8	25
2. *Falda*			
weaving	5–8 days	2-1/2L. = 20	55–60
dyeing	1 week		
sewing	1 hour		2
3. *White gauze sábana*			
weaving	2–3 days	1/2L. = 4	14–16
bleach	1 day		.5
punto	2 days		4–8

TABLE 1 *Continued*

	Objects & Processes	*Time of Producing*	*Material Cost in Pesos*	*Selling Price in Pesos*
3a.	*Gauze with colors*			
	weaving	3–4 days	1/2L. = 4	25–30
	punto	2 days	24 C = 14.4	5–8
4.	*Rayes sábana*			
	weaving	4–5 days	1 L. = 8	35–40
	punto	2 days		5–8
5.	*Panal de Abeja Sábana*			
	weaving	4–5 days	1 L. = 8	35–40
	punto	2 days		5–9
6.	*Cinedor*			
	weaving	3 days	2/3L. = 6	20–25
	punto	2 days		5–8

L. = libra (pound) C = Colores (package of colored thread)

Note: Prices recorded here are averages. Time is estimated, based on observation and data collected from producers, and intended to provide relative time differences for producing the objects.

The other principals in the economic system are: 1) the patron-buyer (Pa), either the head of a putting-out system or a contract buyer, is a local woman, who buys to sell products in Oaxaca and locally only if the opportunity arises; she either contracts out the finishing processes (FP), or does them herself. 2) The *tienda*-keeper (T) a type of patron, buys on contract, or for repayment of credit and almost always contracts for the finishing processes. Tienda-keepers sell to vendors, to occasional tourists (N-pR,T), or other visiting strangers. 3) A vendor (V), is a trader, a non-resident man, who travels to Yalálag's weekly market to sell outside wares and to buy local products to sell in Oaxaca and elsewhere. 4) An embroiderer (E) is a woman who knows how to embroider, but whose work is either not known, or not good enough to be contracted for. 5) An individual (I) is a weaver's neighbor or relative, who buys a garment directly for personal use.

The finishing processes (FP) of sewing and hemming, *punto*, embroidery, and bleaching, are usually done on contract for patrons and *tienda*-keepers and similarly to finish the personal garments of weavers and individuals.

When the objects are completed, some, as noted above, are sold locally by two or three patrons and *tienda*-keepers, but the bulk, amounting to three quarters of the output, particularly white gauze *sábanas* and costume-*huipiles*, are carried to Oaxaca, (final sales, FS) to be sold to Oaxaca *tiendas* (OT) or in the Oaxaca market (OM). In addition, but very infrequently, vendors sell a few textiles in other village markets (OVM). Finally dealers (D), buy textiles from Oaxaca outlets to sell (extended sales, ES) elsewhere in Mexico (Mexico stores, MS) or to ship to the United States (United States stores, U.S.S.).

Supply and demand more or less control the sale and distribution of the textiles produced for local Yalálag use, in contrast to the Oaxaca marketplace, where neither the quantity, quality, nor variety of textiles meets the potential demand. Supply is set by the producers, a situation arising, in part, from their misapprehension of the marketplace and their adherence to convention more than a desire to exert control. The majority of weavers and patrons accept the general notion that tourists or the Oaxaca marketplace are interested in buying only costume *huipiles* and gauze sábanas. Like traditional artists in general, who believe that low quality is "good enough for tourists," they export inferior textiles (see also Crowley 1974; Pi-Sunyer 1973). With the supply thus artificially created, tourists, unaware of possible alternatives, buy what is on hand. These aspects of marketing, as well as other elements of the economic system influence not only the weavers' decisions of what to weave, but how well to weave.

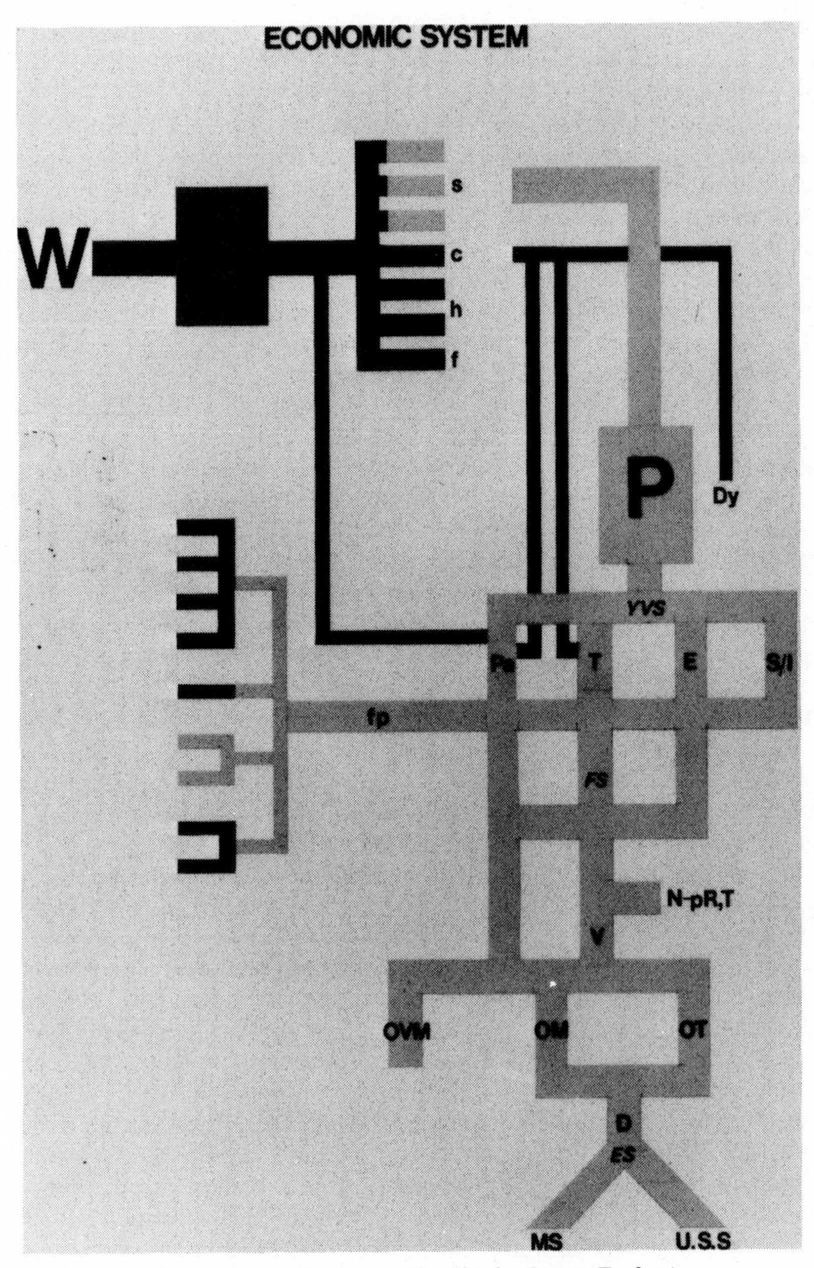

Figure 10. Flow chart of economic system. (Graphic by Jeanne Taylor.)

Decision Process

Returning to the weavers' decision process, we can begin with their economic situation to sort out the determining factors. The immediate subsistence requirements of the most impoverished, who, for example, need money for food daily, may force them to choose to weave a gauze *sábana* (s) in a period of three days at a low return of four to five pesos daily rather than earning seven to ten pesos daily for a *huipil* (h) which requires a week to make (see Table 1). Furthermore, the most impoverished are poor candidates for credit at the *tienda* and, to minimize risk, usually find it necessary to work as members of a putting-out system where the patron decides the type of textiles to be woven. Since a weaver is paid only for her labor in such cases, and is regarded as a *pura esclava* (mere slave) it is unlikely that she is inspired to do more than *corriente* work. *Corriente* gauze *sábanas* are, in fact, an economically and technologically satisfactory solution for these weavers; they can be made rapidly, cost less in time and materials, have a ready market, and their loose, light weave calls for only moderate physical effort. Further, the very small differential earnings for a finer product are not worth the necessary investment in time and energy, especially when the product is to be exported and sold to remote individuals who are assumed not to know the difference. Moreover, gauze *sábanas* are desirable products from the point of view of many patrons, often poor themselves, who are oriented toward quantity sales. The large supply of inferior gauze *sábanas* produced results primarily from the weavers' and patrons' economic situations combined with these other factors.

When their economic situation or patron relationships permit, these same weavers and others can choose to increase their earning capacity by making *huipiles* and other products. Weavers who are economically secure, either because of weaving ability or supplementary sources of subsistence can use finer materials, earn slightly more and add to their prestige by selling their work to superior patrons who demand finer quality. They can exert more control over the type and quantity of objects woven. Indeed, their aesthetic motivations or talent may lead them to choose to weave for their personal satisfaction as well as for economic gain. Some successful weavers spend all their time weaving, depending almost entirely on weaving earnings, while others weave only occasionally to supplement other sources of income. Both types have stable patron relationships and are contracted for a variety of textiles, some of which are technologically difficult, and probably also aesthetically rewarding.

To some extent, each weaver's choice of action is based on current market demands combined with her immediate personal needs, and seem to be ad hoc (Nash 1961), but there are certain regularities in the economic system. All weavers, regardless of ability or economic status make their best textiles, first for themselves and second to sell to individuals in Yalálag for local use. These products, by no means all of them *fino* in quality, advertise weaving ability, thereby adding a weaver's prestige and her possible economic gain. The finest weaving produced for sale is contracted for, with transaction initiated by the patron-buyer; included here, also, are embroiderers and *tienda*-keepers. Objects peddled by the weavers themselves are usually inferior.

Certain weavers, regardless of their economic situation have the capacity to create *fino* textiles, and some are motivated to make them to sell. By returning to the aesthetic characteristics of the products, we can attempt to assemble the factors necessary for the creation of fine quality textiles.

FACTORS ESSENTIAL TO FINE WEAVING

First, *fino* weaving is made by weavers who have the requisite aesthetic ability and technological capacity. With sufficient talent and dedication, even the most impoverished can eventually build a reputation and overcome economic constraints.

Second, *fino* weaving is commonly available only from superior patrons (Fig. 11) who seek out superior weavers, respect their abilities and reward them in diverse ways: by regular buying habits, by advising about quality—assisting in color selection, providing pattern direction, encouraging the use of fine thread—and by prompt and adequate pay. Patrons are key figures in the system and their aesthetic, technological, and economic contributions should not be underestimated. Superior patrons have developed aesthetic sense which enables them to identify the products of the best weavers. In addition, they make specific recommendations with respect to color, in particular, and also weaving technique. Moreover, they are aware of the complexities of the Oaxaca marketplace. Knowing that certain stores pay more for quality work and seeing their own reputations at stake, they encourage fine weaving, *punto,* and embroidery. In contrast, inferior patrons, who are less aesthetic, or less affluent, or less knowledgeable, rely on quantity transactions and sell inferior textiles in the Oaxaca market at low prices. Although association with a superior patron enhances the prestige of a weaver, only those who are strongly motivated as well as highly qualified risk the opportunity. These weavers are rewarded economically by secure income and psychologically by recognition and the pleasure of doing fine work.

In addition to these rewards, motivation is affected by the perceptions of the costume-*huipil,* that is seen either as symbol and inspiration to all Yalálag, or as a locally created beautiful object, which, along with other textiles, is bought by people from all over the world. Certain older weavers, who take pride in their work, have been able to see the modern response to the costume-*huipil* as an extension of their own traditional attitudes, while the perceptions of some patrons have shifted and they see it in contemporary terms as handcrafted folk art, something of aesthetic value which is admired for its artistic excellence by discerning tourists. Both views encourage fine weaving.

Figure 11. Yalálag's most influential and demanding patron whose weaving products are sought as presentation gifts for important visitors, as bridal gifts, and by occasional visiting tourists. She is pictured in her patio doing embroidery at which she excels.

Regular or average weaving is made by weavers who are incapable or unwilling to meet the perfectionist standards of the best patrons. *Corriente* or poor weaving, made intentionally, is primarily resorted to as a means of subsistence in which aesthetics plays no part. Young, dress-wearing weavers, in particular, subscribing to the school-indoctrinated attitudes about the backwardness of wearing the *huipil* and unable to perceive their work as folk art, resent the circumstances that force them to weave, and consequently make shoddy textiles.

CONCLUSIONS

The weaving complex has been adaptive for Yalálag, providing a means of support to a range of both weavers and others of varying capacities and situations. It distributes risk; no one gains very much, but no one loses. Fitting Yalálag needs, it encourages and supports differential weaving quality rather than the quantity production of fine weaving. The weaving complex is also adaptive for individuals, providing sufficient flexibility for weavers to resolve their varying personal needs (Alland 1973).

The basic structure of the weaving complex evidently has remained unchanged; only the weaving of garments for export has replaced the production of *manta.* The efficient flexibility of the system will probably continue as long as weaving endures, although it is not an actual reflection of the marketplace. Even if the price structure were altered drastically to encourage more quality weaving, it seems doubtful that many women would be attracted to it. Emigration, machine-sewing, and other preferred means of subsistence, as well as the attrition of older weavers are contributing to the demise of the craft, although it is not possible to predict its disappearance with certainty, since many apparently moribund traditional arts have been revived by the current interest in folk art (Crowley 1974).

The total weaving system combining all the above aesthetic, technological, and economic factors as well as many other social factors not touched upon here, is responsible for both the character of the weaving and its continuation. This leads one to question the assumptions of the aficionados and collectors of folk/traditional art, who generally believe that direct contact with artists and the elimination of middle men is more conducive to stimulating an enduring fine craft. Clearly, that type of shift to greater personalization, part of a general shift from traditional to folk art, changes the art as the system is modified (Paz 1974). Such a trend seems only minimally underway in Yalálag, where patrons still hold sway even though the best weavers can be identified with their own creations.

LITERATURE CITED

Alland, Alexander, Jr. and Bonnie McCay

1973 The Concept of Adaptation in Biological and Cultural Evolution. *In* Handbook of Social and Cultural Anthropology. John J. Honigmann, ed. Pp. 143–178. Chicago: Rand McNally.

Crowley, Daniel J.

1974 The West African Art Market Revisited. African Arts 7 (4):54–9.

Edwards, J. M. B.

1968 Creativity: Social Aspects. *In* International Encyclopedia of the Social Sciences. Vol. 15. Pp. 442–57. New York: Macmillan.

Jopling, Carol F.

1973 Women Weavers of Yalálag; Their Art and Its Process. Unpublished Ph.D. dissertation, University of Massachusetts, Amherst.

Leene, Jentina E.

1967 Some Remarks on Chemical Analysis of Ancient Textiles. *In* Application of Science in Examination of Works of Art. Pp. 238–45. Proceedings of the Seminar, September 7–16, 1965, conducted by the Research Laboratory, Museum of Fine Arts, Boston, Mass.

Merrill, Robert S.

1968 Technology. *In* International Encyclopedia of the Social Sciences, Vol. 15. Pp. 576–89.

Nash, Manning

1961 The Social Context of Economic Choice in a Small Society. Man 219:186–91.

Ong, Walter J., S. J.

1969 World as View and World as Event. American Anthropologist 71:634–47.

Paz, Octavio

1974 Use and Contemplation. *In* In Praise of Hands; Contemporary Crafts of the World. Pp. 17–24. Greenwich, Conn.: New York Graphic Society.

Pi-Sunyer, Oriol

1973 Tourism and Its Discontents: the Impact of a New Industry on a Catalan Community. Studies in European Society 1:1–20.

†